The Complete Guide To IT

Service Level Agreements

Aligning IT Services to Business Needs

◆

THIRD EDITION

◆

Andrew Hiles

ISBN 1-931332-13-4

The Rothstein Catalog On Service Level Books
Rothstein Associates Inc.
Brookfield, Connecticut USA
www.ServiceLevelBooks.com

Third Edition
by Andrew Hiles

Third Edition, Published January, 2002.
Second Edition, Published 2000.
First Edition, Published 1991.

ISBN #1-931332-13-4

Although great care has been taken in the compilation and preparation of this book to ensure accuracy, the author and publishers cannot under any circumstances accept responsibility for any errors or omissions. Service Level Agreements is a fast-moving subject area and inevitably some of the information contained in the book will become out of date. Readers should verify any information contained in this book before making any commitments.

No responsibility is assumed by the Publisher or Author for any injury and/or damage to persons or property as a matter of product liability, negligence or otherwise, or from any use or operation of any methods, products, instructions or ideas contained in the material herein.

Readers should be aware that International, Federal, National, State and Local laws and standards may affect the contents and advice contained in this work, which should be interpreted accordingly. Only the courts can authoritatively interpret the law.

The author welcomes suggestions for inclusion or improvements.

PUBLISHER

The Rothstein Catalog On Service Level Books
Rothstein Associates Inc.
4 Arapaho Road
Brookfield
Connecticut 0608-3104
USA
☎ 203.740.7444
☎ 1-888.ROTHSTEin
203.740.7401 fax
www.rothstein.com
www.ServiceLevelBooks.com
info@rothstein.com

AUTHOR
Andrew Hiles

Kingswell International Limited

The Forge
Kingston Bagpuize
Oxon OX13 5AG
United Kingdom
+44 (0) 1865 822010 ☎

+44 (0) 1865 822011 fax
www.kingswell.net

info@kingswell.net

OTHER BOOKS BY THE AUTHOR

Hiles A.N., **SLA Framework CD-ROM**, 2001, published by Rothstein Associates Inc., ISBN #0-641648-5-X. This CD-ROM is the companion to this Guide, and provides extensive sample Service Level Agreements from a broad variety of business and technology environments.

Service Level Agreements, Winning a Competitive Edge for Supply and Support Services. Revised and enlarged 2000 edition, ISBN 0-9641648-4-1, published by Rothstein Associates Inc. This book applies Service Level Agreements to services other than IT. Real case studies and example SLAs are provided ranging from Human Resources, Logistics, through Training, Livestock Handling, Logistics and Field Service Engineering.

Creating a Customer-Focused Help Desk: How to Win and Keep Your Customers, with Dr. Yvonne Gunn. Published by Rothstein Associates Inc. 2000, ISBN 0-9641648-6-8. This book has the support of the Help Desk Institute. www.helpdeskinst.com.

Hiles A.N.: **Business Continuity: Best Practices**. Published by Rothstein Associates Inc. 2000, ISBN 0-9641648-3-3, updated 2001. This book explicitly covers all the ten areas of business continuity competence required for membership of the Disaster Recovery Institute International (DRII) and the Business Continuity Institute (BCI).

Hiles, A.N.: **Enterprise Risk Assessment and Business Impact Analysis: Best Practices**. 2002, published by Rothstein Associates Inc., ISBN #1-931332-12-6.

Hiles A.N. and Barnes, P. (Editor and main contributor) **The Definitive Handbook of Business Continuity Management**, John Wiley & Sons, 1999, ISBN 0 471 98622 4. This book is produced by Survive, the international user group for business continuity.

Hiles, A.N.: **E-Business Service Level Agreements: Strategies for Service Providers, E-Commerce and Outsourcing.** 2002, ISBN #0-9641648-9-2. Published by Rothstein Associates Inc.

All of the above books may be obtained from:
ROTHSTEIN ASSOCIATES INC.
4 Arapaho Road, Brookfield, Connecticut 0608-3104 USA
www.rothstein.com or www.ServiceLevelBooks.com
e-mail info@rothstein.com
Phone (USA): 1.888.ROTHSTEin; (Worldwide) 203.740.7444; Fax: 203.740.7401

Mr. Hiles has also contributed to:

- **Croner's Purchasing and Supply Guide to I.T.**, 1994, ISBN 1 85524 271 0
- **Guide to Business Continuity Management**, 1999, for the Confederation of British Industry by Caspian Publishing
- **Business Continuity Management**, 2000, Institute of Directors / Department of Trade & Industry.

ACKNOWLEDGEMENTS

Acknowledgements are due to:

Mr. Bill Barham, Lloyds Bank Plc, for his input into the format of the standard SLA.

Dr. Brian Day, past Performance Manager of the Joint European Torus Computing Centre and Harwell Computing Centre, past Vice Chairman of the UK Computer Measurement Group, for his suggestions and advice.

Mr. George W. A. (Bill) Miller, American Airlines, for his excellent SLA Checklist which has been adapted in this book.

CONTENTS

LIST OF FIGURES

FOREWORD

Most suppliers lose around 16% of their customers each year. The reason? Poor service. Typically if you provide good service, your customer may tell five people. "Customer promiscuity" is the norm: your customers and prospects are one click away from your competitors. Discontented customers typically tell over *ten* people how bad you are. In the days of bulletin boards, a discontented customer can place messages that can impact - maybe even destroy - your business. In the often dangerous and unpredictable e-world, Service Level Agreements are imperative to protect both parties.

Any support service, whether in-house, contracted or outsourced, stands to be accused of being insensitive to the requirements of its customers (or users). Equally, customers of a support service may have unrealistic expectations of what can be reasonably provided by it. Service Level Agreements can overcome these gulfs.

All too often service level reports are misleading: bad statistics, measured in ways and at points that do not truly reflect the service experience of the customer. This book exposes pitfalls, problems and challenges in e-business Service Level Agreements and lays the foundation for harmonious and effective customer-supplier relationships to enable actual service delivery to become aligned to customer expectations.

What, then, is a Service Level Agreement? A Service Level Agreement is simply an agreement between the support service and the user quantifying the minimum acceptable service to the user. SLAs are particularly valuable in real time activities of e-commerce where speed-to-market is crucial; where there is no time for mistakes; and where millions of dollars can be lost in minutes.

A Service Level Agreement can create harmony between the parties, and avoid disputes between customer and supplier. It can justify investment and identify the "right" quality of service. It can mean the difference between business success and failure.

Dr Yvonne Gunn
Kingswell International

PREFACE TO THE THIRD EDITION

Some 15 years ago, the UK IBM Guide Operations Managers group (of which I was Chair) held a meeting in which we discussed the concept of SLAs. I had read about the pioneering work in this area by Bill Miller of American Airlines and developed by the Capacity Management Group. Seizing on the concept as good management practice, I first implemented them in the company for which I was then working. It really was leading edge stuff in those days and there was little guidance, so in 1988 I began to present training workshops on the topic. These aroused considerable interest and I was persuaded to write the first book on the subject, published by Elsevier. Believing this concept was equally applicable to any support or supply service, the second book, suggesting this transition, quickly followed. We began to receive inquiries about SLAs from a wide range of public and private sector enterprises, covering a broad spread of business and support functions. Since then we have presented on SLAs at conferences and workshops around the world and written literally hundreds of articles on the subject.

My vision in SLAs is simple: SLAs are potentially a strategic tool to align all support services (especially IT) directly to business mission achievement. Sadly, few organizations use them in this way.

The early SLAs were IT-centric, written in IT technical terms, and predominantly provided the IT user with service levels that had more to do with internal IT performance measurements than with business-oriented service achievement. Frequently metrics were inappropriate, measurements imprecise and monitoring weak. The SLA reports simply did not reflect the experience of the customer when using the service. Now, the more mature organization writes business-centric SLAs and has sophisticated performance measurement tools that accurately reflect the customer's or service user's actual experience. Unfortunately, we are now seeing the legal profession moving into the field of drafting SLAs, changing the concept from a crystal-clear definition of the service and of service levels, back into a muddy, legalistic puddle.

Back full circle? It's time to start over.

That is why this book is particularly important today. We must keep the service vision, definition and requirements clear – even more important today in a time of loose partnerships, complex – often virtual – supply chains and instant success or failure. In the e-world, particularly, customers are just one click from desertion.

Where are SLAs going? Increasingly business-focused. Increasingly measured in real-time. Simple documents that cover complex service infrastructures. Providing competitive edge. Embracing penalties. The brave, who commit to tight SLAs and perform against them will win the commercial spoils.

Andrew Hiles

Oxon Bagpuize, England

1

AN OVERVIEW OF SERVICE LEVEL AGREEMENTS: WHAT THEY CAN AND CANNOT DO

1.1 Introduction

A sign of a maturing technology is that it gets so deeply interwoven into the fabric of business, industry, government and society that existence without it becomes inconceivable. Computing systems and telephony, data communication and Internet based services have permeated virtually every aspect of life and, consciously or not, society now relies on them. Air, road, rail and maritime control systems; banking and finance; wholesale and distribution; transport; retail; health; leisure; manufacturing industry; government; communications and the media - all are now as reliant upon these services as they are upon electricity, water or fuel. Enterprise Resource Planning (ERP) and Customer Relationship Management (CRM) systems create inter-related dependencies throughout the organization, while the growth of Electronic Data Interchange (EDI) and e-business extend reliance on computing from corporate dependence to inter-corporate dependence.

The trend towards outsourcing has exacerbated supplier dependence. Over 50% of outsourcing contracts involve dispute at some stage – frequently ending in changing the supplier with consequent hiatus to both businesses. The cause is usually weak contracts or poor service specifications: how can we avoid such damaging attrition?

Call Centers have been one of the fastest growing areas in technology over the last few years. Many Call Center operations are outsourced. If the Call Center is unavailable, or if response is slow, customers simply go elsewhere – to the competition. High availability and quick response are vital for customer gain and retention.

The rise of e-business adds to the increasing chain of interdependencies and, even more importantly, to the speed and concatenation of the impact of loss of service or poor service. Reliance on Internet Service Providers, Application Service Providers, Managed Service Providers and Total Service Providers and all their intermediate suppliers means that a failure in any link in this supplier chain becomes a failure of the whole chain, with potentially disastrous impact: one bank has claimed that failure of their ISP service could potentially cost $1B in an hour.

Yesterday's leading edge becomes today's utility and tomorrow's passé obsolescence. With Information Technology (IT) we are constantly facing a dichotomy: how do we manage the "bleeding edge", transformational technology as well as the well-established IT utility services?

The Utility Service

Many computing services are being seen as utility services just like electricity or water. These days, utilities worldwide have increasingly become privatized - expected to fund themselves as commercial entities rather than be provided by governments. Similarly, in-house computing services are increasingly expected to be self-sufficient, at least recovering costs from customers, rather than being provided as a corporate service as part of the overhead costs.

As with the supply of any utility, the end user expects a defect-free service, available 100% of the time, at a reasonable cost. A utility service implies a limited range of standard products - but this is in direct conflict with the increasing technological complexity with which the Information Technology Manager has to cope and the greatly increasing range of services supported . especially common user services like electronic mail, office systems, large databases and decision support systems. The Information Technology Manager is often trying to cover too much ground with too few resources: he or she needs to standardise and prioritise service offerings.

All too often, however, this logic is not extended to the provision of the Information Technology service. You know what the tariff rate is for power, gas, fuel oil and all the other utilities. The price may depend upon your negotiating power . but you know what you are getting for your money. You can measure the consumption of these services in terms of efficiency and select the appropriate quality: premium, unleaded or 4 star gasoline for instance. Your organization specifies the quality it requires for any consumable that it uses. This makes sense: buying goods of a higher quality than you need just wastes money.

Again, this utility logic may not have been applied to the Information Technology services. A specification has probably been drawn up on a project by project basis to establish the amount of storage and computer power required. Throughput requirements may also have been stated . at least for newer systems and especially for transaction processing systems. But there are many other aspects of the services that frequently are not quantified. Often the service requirement has not been recently checked back with the user so changing user requirements may not have been expressed in the service specification.

All this can lead to a mismatch of expectations between the end user and the service provider. The end user perceives that the service is 'poor.' Response may be 'slow.' Support may be 'patchy.' All these are unquantified, but they suggest a background noise of end user dissatisfaction with the service, especially vocal if the end user is a paying customer. It is all too easy for the Information Technology Manager to assume the user's perception is wrong: it cannot be wrong! It is the user's perception and will remain so until it is changed.

The Information Technology Manager, being frequently more analytical than the end user, will doubtless be measuring certain aspects of the service which are perceived as key performance indicators. As long as these are at worst consistent and at best improving, the Information Technology Manager may believe that a good service is being provided.

This service may not however be what the customer wants! So how do we align the Information Technology service to the needs of its customers?

The answer lies in Service Level Agreements.

Leading Edge Services

Leading edge services are frequently high risk and may rely on:

- Unstable equipment
- Unproven and potentially error-prone applications
- A creaking infrastructure
- Links to and use off legacy systems that were not designed for the new usage
- New and possibly inexperienced suppliers.

They are often seen as high risk, high reward, with the prize going to the first to market. Often, service management disciplines are jettisoned in the rush to be first: in most cases the result is failure and the prize goes to the tortoise who came second or third rather than the hare that came first.

With these services, it is crucial to get the service parameters right and to ensure the supplier meets deadlines, timescales, availability and response requirements.

The answer to these aspects also lies in Service Level Agreements.

1.2 Service Level Agreements: Definition

Many contracts are weak in service definition and many in-company services have unspecified or poorly specified service products and service commitments. Service Level Agreements fully define the services to be supplied and the service quality and timeliness or responsiveness associated with them in unambiguous terms.

For internal service providers, this provides both a statement of what they do and the quality of the delivered service and so can provide justification for budget and resource. Where several internal departments are involved with the end-to-end delivery of a service, SLAs[1] between those departments ensure that the customer-facing department is able to enter into an end-to-end SLA with the customer on the basis of the commitments made to it by the other internal departments.

Commercial services would normally be provided under a contract. However, many contracts for external services are vague, and weak in defining the service and the quality and timeliness aspects of it. Moreover, frequently service aspects are littered amongst legal and commercial sections so it is difficult to get an overall view of the service and its quality. In this case, a Service Level Agreement may supplement a contract. For commercial suppliers, a SLA identifies, in one place, the tasks, the deliverables and the quality and time aspects of the service. It also limits their exposure: if it is not in the contract or SLA, it is a chargeable extra.

[1] These inter-departmental SLAs are sometimes called Operational Level Agreements (OLAs)

Where several suppliers are involved in the end-to-end delivery of a service, back-to-back SLAs[2] are necessary so the lead supplier can provide an end-to-end SLA to the customer.

A Service Level Agreement, is simply:

> *"An agreement between the computing service provider and the user quantifying the minimum acceptable service to the user."*

Each of those words is important. A **Service Level Agreement** (SLA) takes the form "if you do this, then I will be able to do that, and this will be the cost." It has to be negotiated and **agreed.** Agreement is two-way. It depends on the user providing accurate utilization and volume forecasts (perhaps not exceeding 'n' transactions per hour) and in the user meeting deadlines (say for completion of input to the system). The agreement usually takes the form: "if you do this, then I will be able to do that." As with any negotiation each side will start out with an opening position. This position will be modified as negotiations continue, to take account of practical difficulty and cost. This negotiation should not be some sort of duel between the IT professional and the user: it should be a joint exploration of what is in the best interests of the customer's business. For in-house computing services it is also their business.

Quantifying the service is vital. There is an old saying "what can't be measured doesn't exist". Quantification of various aspects of the service will place assessment of its quality on an objective basis and get away from these vague, emotive words like "slow" and "poor."

The service specification should be the minimum service level that the user is prepared to accept. The end users' requirement will be examined to establish what the benefits of various levels of service are and what level is cost justified. There is probably no benefit in over-providing quality: it just

[2] These back-to-back SLAs may also be known as tiered SLAs or multi-tiered SLAs.

costs more money. Over provision of quality for instance by providing sub-second response time - when there is spare capacity may just raise expectations to a level that becomes unsupportable later when the workload increases and that initial quality cannot be maintained. The service has to be fully adequate but it need be no more than that. However, the quality must also be consistent since inconsistent quality is equated to poor quality by the customer.

The definition of **minimum** (that is, adequate) has to be explored. The pyramid model of service and value (Figure 1) may help in this process. This model identifies four levels of service from Level Zero (the end-users are on their own, without the Information Technology service support) to Level Four (highest quality service).

Figure 1.1: Service Support Characteristics

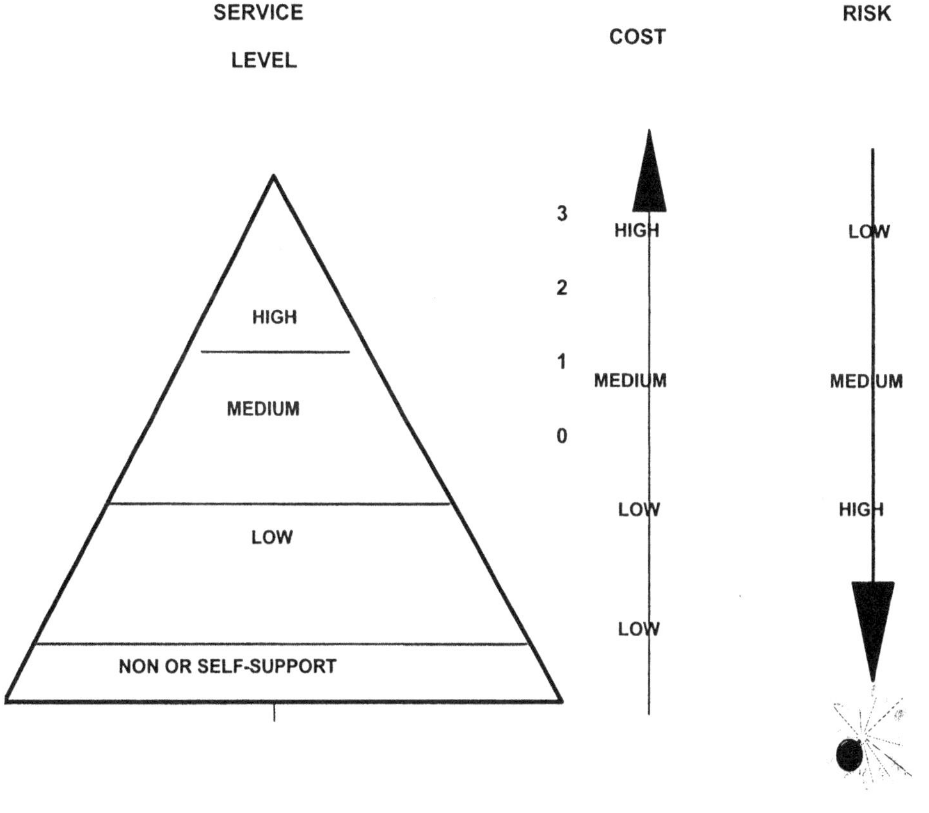

Level Zero may apply to turnkey departmental systems; to systems where user expertise exceeds that of the computing service; or to non-critical systems. Level Three may apply to mission critical or high value financial systems. In general, the higher the level of service, the higher is the service cost: the lower the level, the greater is the risk of outage. In presenting multiple options for levels of service, management is prompted to establish the quality of service appropriate to that system. This process defines the **minimum** service requirement.

Thus if the Information Technology service consistently meets the minimum standards its service will be perceived by the end user as good. If it over-achieves that does not necessarily mean a better service: it may just mean a more expensive service.

Minimum also applies to the range of services offered. Does the user need a Half-dozen different databases, five different spreadsheets, three different project management packages (as well as the project accounting facilities in the corporate accounting software)? The more software with which we clog our systems, the more the service costs - in capacity, in software purchase, annual licence and maintenance costs and in support. So we need to establish the minimum range of facilities we should be offering.

The service has to be **acceptable to the user.** Users have their job to do and IT only exists to help them to do it. Many companies arc undergoing a fundamental change of perception about their internal Service Providers and Business Units. Particularly if the service is charged out, "customer" is a more appropriate word. In this case, the SLA may also need to contain charge-out terms. In any event, users are our customers - so let us treat them as such.

That is another benefit of a SLA: it promotes customer orientation within the:

- mission statements
- strategy
- 5 year plans
- 3 year plans
- business analysis
- system specification

- departmental annual targets
- individual objectives for their staff.

Departments and staff can then deliver their contribution to the corporate mission. Corporate achievement of its targets is thus a sort of project, with a series of deliverables against deadlines and milestones. Each department (and each member of staff) has to be considered "contracted" to produce specified deliverables if the corporate plan is to be achieved. Where these departments rely on an Information Technology service, the service should be "sub-contracted" to provide computing of appropriate capacity, quality and cost. A SLA represents this "sub-contract" with an in-house Information Technology service or amplifies a contract for an external service.

A SLA therefore orients the Information Technology service to the business plan - and pre-empts situations where technology is acquired for its own sake or an IT empire is allowed to grow up without quantification of its real contribution to the business.

1.3 Serving the Business

Every organization has a mission. Public Authorities exist to provide certain defined services and their political masters may dictate the quality and cost criteria to which those services are to be provided. Commercial undertakings exist to provide profits for their shareholders by providing defined products or services. Every successful business has a clear understanding of its role, often defined in a mission statement. This is backed up by a defined corporate strategy, by a five-year plan, a three-year plan and detailed annual business targets. The corporate plan is supported by business analysis that specifies IT services and by senior management setting departmental objectives. Individual departments within the organization then set their own:

- mission statements
- strategy
- 5 year plans

- 3 year plans
- business analysis
- system specification
- departmental annual targets
- individual objectives for their staff.

Departments and staff can then deliver their contribution to the corporate mission. Corporate achievement of its targets is thus a sort of project, with a series of deliverables against deadlines and milestones. Each department (and each member of staff) has to be considered "contracted" to produce specified deliverables if the corporate plan is to be achieved. Where these departments rely on an Information Technology service, the service should be "sub-contracted" to provide computing and communications services of appropriate capacity, quality and cost. A SLA represents this "sub-contract" with an in-house service or amplifies a contract for an external service.

Service quality should reflect business objectives and priorities. The Mission and vision of the organization (e.g. "to be number one") are subjected to business analysis to establish what has to take place for the mission to be achieved. These achievements can be expressed in Critical Success Factors (e.g. "to ensure continuity of supply to customers") and in specifications for new projects (e.g. "a new, quicker and more robust order processing system"). High Level Key Performance Indicators or Key Ratios are developed to track whether the organization is maintaining achievement of these Critical Success Factors (e.g. "items go from order to delivery within 24 hours"). These Key Performance Indicators can then be decomposed to reveal the processes and activities on which their achievement depends (e.g. "no more than 5% of late orders"). Service Levels can then be applied to these processes and to the suppliers, equipment and infrastructure that supports them. (e.g. "the Application Service Provider must have a Service Level for the Order processing application of 99.99% availability to the customer organization"). A schematic of the way in which this is achieved is at Figure 1.2.

Figure 1.2: Serving the Business

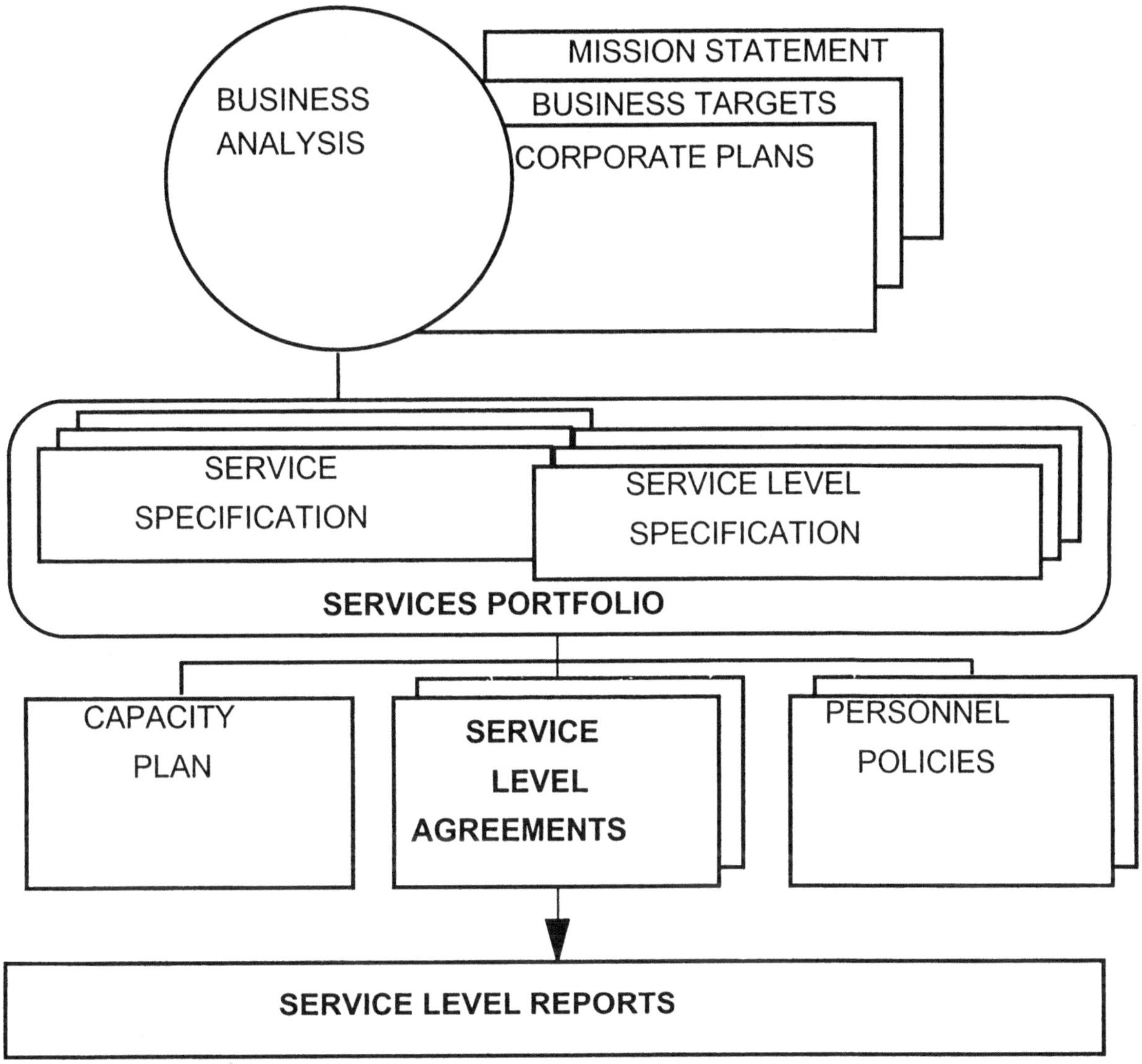

1.4 Availability

Service Level Agreements work to identify appropriate service parameters for the computing service. The most obvious of these parameters is service availability.

What do we mean by availability? Is it simply the uptime of the CPU? Or do we mean the total system . do we include data storage devices, printers, plotters and other peripherals? And is availability defined solely at the computing service, or do we include all telecommunications links, intermediate controllers, modems, terminals, ATMs, EPOS equipment and remote printers? What about distributed processors connected to a central system: should they be covered by a SLA? And microcomputers, desktops, workstations, Local Area Networks (LANs): should the SLA include them, too? Do we measure availability at the computing service, or at the end-user's desk?

In short, do we include all devices and **all** access paths used within the organization, or just those judged to be critical? Where alternative paths or devices exist, should availability be taken to mean the availability of at least one device or one path? Similarly, in a multi-CPU environment, is the computer "available" despite one or more CPUs being out of service?

And what about downstream end-users, accessing our computer systems via computers which are part of other external networks: they may complain of our unavailability, but that unavailability may be due to malfunction of a device within the external network.

In a 'utility' environment, the supplier usually has well defined "end points" or "service points". SLAs should define these.

Even if all the hardware and all the telecommunications links are running, there may be a software problem impacting the service which prevents all or some users from using it. At what stage does that fault make the service "unavailable?" If one user is impacted? If ten are impacted? If more than half the user population is? What if the user can access some, but not all, of their applications or services?

Reliability can be considered part of availability: an availability target expressed in terms of percentage uptime could conceal poor reliability with numerous short breaks. The number and duration of outages is therefore also a significant indicator of the quality of availability.

A definition of "availability" that is valid for any particular computing service, or for a particular group of customers, can only be established by exploring these issues with customers. In general, however, the customer tends increasingly to be interested in availability on his desk. As far as the customer is concerned, the system is unavailable if he or she cannot access it: and the actual location or cause of the fault does not affect this perception!

The corollary of this is that, if the customer has not wanted to access the service, he doesn't perhaps care whether it was available or not. So availability also needs to be defined in terms of "when." A lower level of availability may be adequate overnight, or at weekends or bank holidays for some services. Availability requirements during normal working hours, prime shift, may be substantially higher than outside normal working hours.

Equally, there may be critical periods where even normal prime shift availability targets are inadequate: for financial systems, near period- or year-end; for salary payrolls at month-end; for retail point-of-sale systems immediately before Christmas and Easter. Critical periods may be different for each system; for each major customer; or for each group of users.

To simplify availability specifications, applications can be grouped into tier ratings. Some examples follow – the exact definitions will depend on the individual organization. Tier 1 may have a 100% availability requirement – mission critical applications with high value. These justify investment in non-stop infrastructure – replication, resilience, 24x365 support. Recovery could be specified at less than one hour.

Tier 2 could be those applications needing, say a 99.8% availability 24x365, with recovery in less than two hours.

Tier 3 could be applications requiring 99.5% availability between 07.00 to 20.00 hours with recovery within one day.

Tier 4 could be applications needing 90% availability with a recovery requirement within three days. More Tiers may be added as necessary.

Figure 1.3 below illustrates this concept.

Figure 1.3: Tiered Availability

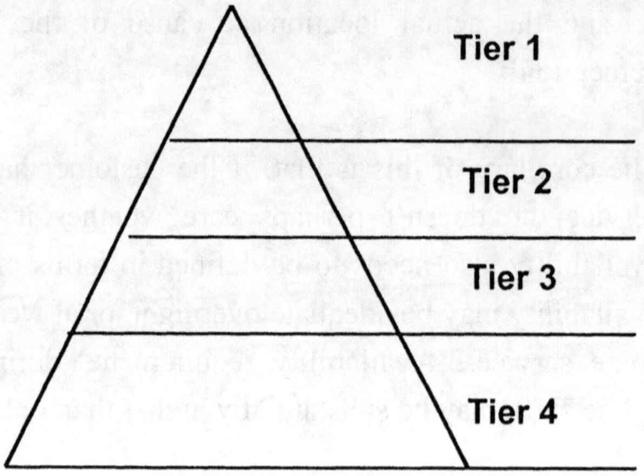

Not just availability, but the appropriate quality of the available service needs to be established.

Having defined "availability" in terms of the components or links and in terms of critical times, other service parameters need addressing.

1.5 Performance: Speed, Response and Accuracy

Job turnaround may be an important factor - input to the system may have to be completed by a specific time and certain jobs may have to be run within a timeframe or complete to meet a deadline. Targets for turnaround and deadline therefore need to be established.

Speed by itself is of little use - speed needs to be accompanied by appropriate accuracy. The old question "Do you want it now, or do you want it right?" is appropriate here: often speed will reduce accuracy, particularly where data is input to a system manually. Accuracy levels therefore also need to be specified - particularly where meeting deadlines and throughput targets may depend on the accuracy of customers' input.

For on-line systems, response is an important quality criterion. Certain types of transactions need to be completed within a defined time. This will vary with the type of application and will depend on whether the user's routine function is dependent on the computing service: a slower response may be acceptable for occasional database enquiries than for a computer-aided design system, for instance.

Response costs money. In the Big Bang (the relaxation of the securities and financial services markets in the UK), stockbrokers Kleinwort Grieveson in the UK specified the maximum response time for 400 trading desks at five seconds. The initial project cost was £12m: to reduce response time to one second would have cost a further £24m!

1.6 Security

Equally, security may be significant. A SLA will need to establish relevant security criteria in terms of physical and logical access to systems, processes, files, data and output as well as providing for disaster recovery and contingency plans.

The SLA therefore has an important role to play in ensuring that service availability and quality is appropriate to business needs.

1.7 Quality

Service quality covers more than just those items mentioned above. Sometimes SLAs have been introduced to enforce compliance with Quality Assurance and Quality Control standards, as part of a zero-defect approach. SLAs are often a key component of Quality Improvement Programs (QIP) and we have implemented them as part of some of the world's largest ISO 9000 and Baldrige Quality programs. If SLAs set high quality standards the computing service has to deliver them or lose credibility.

1.8 Service Culture

The responsiveness of the central data processing functions has typically lagged behind the needs of the end-user. Historically, there have been applications development backlog mountains where worthy but relatively small payback systems have been pushed to the bottom of the priority pile. Some central DP functions got the reputation of telling the end-users what was good for them and generally treating them in a cavalier manner. Others tended to be technology led rather than solution led .perhaps buying "latest and greatest" hardware when the business need was for something cheaper and less sophisticated. In short we have improved, but expectations have improved more quickly.

Yet other computing services have never declared service targets. They have missed deadlines and targets required (but perhaps not specified) by end users. Thus computing services have consequently suffered a reputation for poor service. This may have been because the two-way nature of the service was not fully appreciated: the service can only deliver against service targets if the end-user delivers against agreed input delivery, accuracy and volume targets.

Where service review meetings took place against this background, they tended to degenerate into bitter and personalised affairs: in the absence of objective performance criteria, individuals became blamed for failing to do their job properly.

It was against this background that SLAs were originally designed. They impose a commitment on both sides - on the computing service to provide a defined service and on the end-user to commit to using that service within defined parameters. This de-personalises any service review meetings and leads to objective assessment of performance by both parties.

Often SLAs have been introduced as a means of forcing service management techniques on a computing service lacking these professional methods.

They have also been seized upon as a tool to encourage - impose even - a service orientation on the central computing service. SLAs facilitate a service culture. They establish targets and commit the computing service and end-users - to deliver them.

This is particularly true when the organization is considering internal charging for computing services: where internal users are, for the first time, having to budget and pay for computing these internal customers have a right to a clear definition of the quality of service they are buying.

1.9 But Why SLAs?

A number of arguments can be brought against SLAs:

- the application was originally subject to a specification. The specification should have laid down service criteria. (True, but how many actually did? Besides, life changes, volumes change, value changes and a historic specification may not be valid a year later)
- our staff are professional: we can motivate them to deliver service without SLAs. (Yes, but don't SLAs help by defining objective targets?)
- SLAs confine service to the minimum. (That is the point of them - to the minimum **acceptable** level)
- you do not need SLAs to introduce charge-out. (No, but the customer is entitled to know exactly what he or she is paying for - and how to measure whether they are getting it')
- if Computer Service, Installation and Infrastructure Management methodologies are in place, who needs SLAs? (OK, so you know you're doing a professional job. How does the customer know?)
- why generate SLA overheads? (Quality and control have a cost. However, the shape of the SLA selected need not be the one with the heaviest overheads - pick the right one for your organization)
- who is going to believe the computing service if they are doing the monitoring of performance against SLAs? (Customers are quite

capable of seeing through phoney figures - and customers should be doing some of the monitoring themselves)

- why not just issue a brochure describing the service? That is what the Electricity and Gas providers do. Fine - but if you cover every application, with service targets, you are effectively writing a SLA without imposing any reciprocal agreement on customers to enable the service to meet those targets. What controls demand? Where practicable, it is better to establish parameters of capacity usage to facilitate capacity planning – but this is not always possible. Electricity, Gas, Water, Telecommunication companies, Internet Service Providers and Application Service Providers all have to face up to the rigours of providing a demand-side service.

None of these arguments destroy the fundamental logic behind SLAs.

SLAs force the computing service to orient itself to providing cost-effective computing solutions to serve business needs and to support the business plan. They impel management decisions on the appropriate quality of service to be provided. They correct ivory tower DF astigmatism, directing the vision away from technological issues and into a service perspective.

SLAs may sound like a universal panacea - but there are a number of things a SLA cannot do. A SLA cannot compensate for inadequate definition of business objectives, nor can it compensate for lack of standards. SLAs cannot substitute for poor customer or IT management. They do not obviate the need for other service management disciplines and tools. They cannot be implemented without cost, nor be implemented without resource. Most importantly, they cannot be effective without commitment from both customer and service provider.

Both parties need to deliver their parts of the SLA - the quality of service can only be as good as the accuracy of the customers forecast usage and the customer's adherence to their part of the Agreement.

To see how well prepared your organization is to implement SLAs, complete the Checklist #1.1.

CHECKLIST #1.1: SERVICE ORIENTATION

YES NO

1. Do formal Service Management/Installation Management procedures exist to manage:
   - Service levels?
   - Capacity?
   - Availability?
   - Performance?
   - Problems?
   - Requests?
   - Change?
   - Operations?
   - Security?
   - Infrastructure?
   - Environment?

2. Are these formally documented and:
   - reviewed periodically (e.g. 6-monthly)?
   - subject to formal change procedure when new applications/systems/hardware are brought in with documentation available Telecommunications and to Operations, Systems, Customer Support management and staff?
   - issued automatically to new incumbents of these posts on arrival?
   - are end-users (customers) aware of them?

3. Do Service Level Agreements exist for all applications?

4. Do they cover all services?

5. And all customers?

6. And all key suppliers?

7. Do these Service Level Agreements specify:
   - customer workload in business units (e.g., number of invoices produced)?
   - forecast changes over the duration of the agreement?

YES NO

- customer workload in terms of performance and capacity management?
- peak workload?
- peak workload constraints to limit peak capacity?
- security?
- disaster recovery?
- key records needed for recovery?
- arrangements for monitoring customer satisfaction?
- reports of actual service delivery against service targets
  - for computing service management
  - for customers

8. Are service level targets included in Project Specification?

9. Are they included in application and hardware contracts?

10. Are they included as part of system acceptance testing?

11. Is there a formal sign-off stage at which the computing service accepts new applications as production systems

12. Is there a formal sign-off stage at which the end-user (customer) accepts that new applications are meeting service level requirements?

13. Is service performance checked against the service requirement specification at Post Implementation Reviews?

14. Are Service Level targets set by the if/MIS/Computer Service provider:
    - for each type of usage (e.g. real-time by transaction complexity, batch turnaround, output etc)?
    - for each regime (e.g. prime, non-prime, weekend)?

15. If so, are they published to their customers?

16. Are customers committed to forecast their utilization?

17. And keep to their forecasts?

YES NO

18. Do Service Level Objectives form part of the Computer Service job descriptions and targets for Management by Objectives (MBO)?

19. Is the computing service charged out?

20. If so:
    - are rebates given to customers for failure to achieve Service Level Objectives?
    - and are any penalties applied if customers exceed utilization forecasts?

21. Has a Customer Satisfaction Survey been issued within the last 6 months?

22. If so, does it give scope for customer response on adequacy of Service Levels?

23. And is the Service Level perceived as:
    - better than 12 months ago?
    - worse than 12 months ago
    - constant?

24. Are Service Level Review Meetings regularly held with customers to review actual achievement against service level targets?

25. Are utilization and performance statistics provided to assist this review?

26. Is performance measured at the customer's terminal?

27. Do Computer Operations and Customer Support staff know the value of their customer's applications:
    - in terms of income to the computing service?
    - in terms of value to and impact on the business in the event of downtime?

YES NO

28. Are the relative priorities of each service or application established:
    - for performance?
    - for recovery?
29. Does each application have a designated and documented owner?
30. Has one person been nominated as responsible for all day-to-day aspects of the running of that application?
31. Do customers have a single point of contact within the computing service for any queries, complaints or problems about the service (e.g. a Help Desk)?
32. Is this single point of contact documented and promulgated to customers?
33. Is there a single point of contact in the customer area for the computing service regarding service queries?
34. Is the single point of contact documented and promulgated within the computing service?
35. Are all customer problems (even "trivial" problems) logged?
36. Is a formal, documented escalation procedure in place for customer problems?
37. And is it promulgated to all relevant computing service staff and to customers?
38. Are computing service responsibilities unambiguously defined, documented and promulgated so that this single point of contact knows who, within the computing service, to approach to resolve any query, complaint or problem which arises?
39. For each application and process:
    - is there an inventory of hardware and software?
    - is there a configuration diagram showing inter-related hardware and software?

YES NO

40. Are customer problems interfaced to computing service problem management procedures for hardware problems, software problems, telecommunications problems etc?

41. Are customer problems interfaced to computing service change management procedures?

42. Is change management procedure extended to all changes including documentation and personnel?

43. Are there back-to-back Service Level Agreements with other internal and external suppliers on whom the computing service relies to provide the service?

44. Are formal quality assurance and quality control procedures in place?

45. Are zero defect goals in-place?

2

THE MEASUREMENT OF SERVICE AVAILABILITY AND QUALITY: KEY METRICS AND TECHNIQUES

2.1 Availability: Optimizing Uptime

To ptimise uptime, we need methodologies in place for the day-to-day running of the computer installation. Methodologies to routinely manage:

Services

- central

- distributed
- PC-based
- interactive
- batch
- network
- support.

Depending on:

Resources

- hardware
- networks
- infrastructure
- plant
- environment/space
- software
- people
- suppliers
- finance.

Against:

Objectives

in terms of:

- availability
- performance
- capacity
- security
- quality.

But availability suffers most not from the routine, but from abnormal events. The two most important disciplines in respect of availability are probably management of:

- change, and
- problems.

Figure 2.1 shows the relationship of infrastructure management methodologies to service level reporting.

Figure 2.1:
Service Level Reports and Service Management

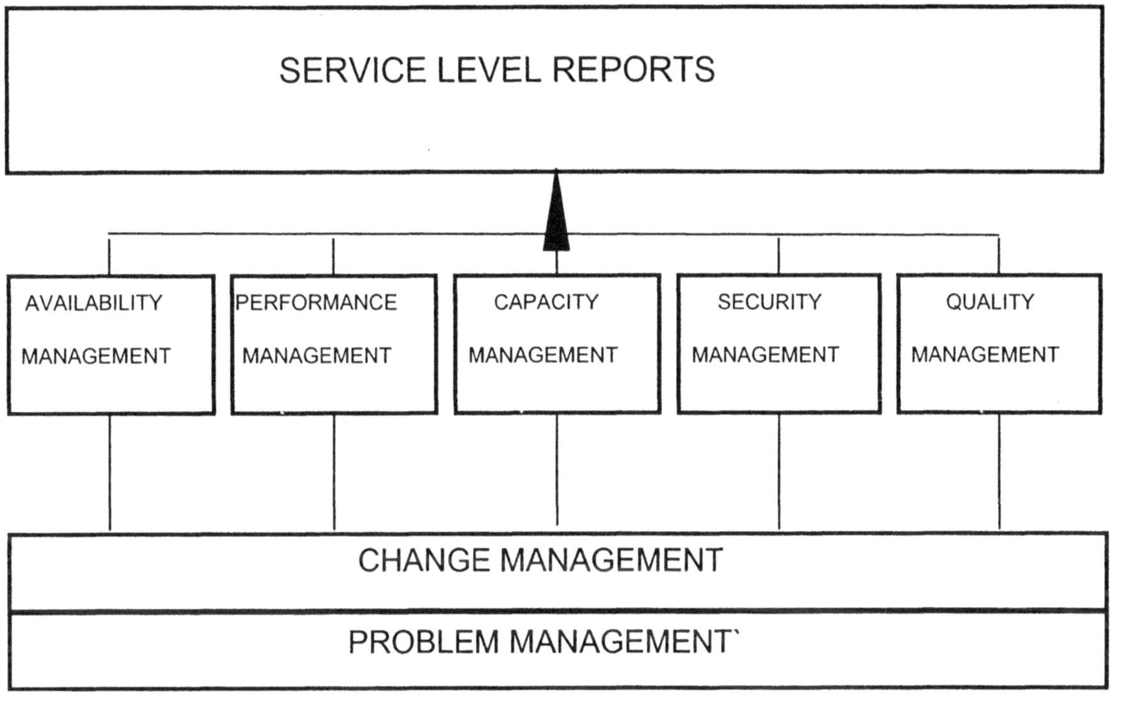

2.2 Change Management

One of the most common causes for lack of systems availability is poor Change Management.

Change Management is normally thought of in a systems or applications context. It is true that increasingly software has taken over from hardware as the prime cause of downtime. In terms of application software, Change

Management needs to control the migration of an application through the various stages of its development into operational acceptance and finally user acceptance. All too often the last two stages are omitted from the implementation process. The penalties for omitting these vital steps are that Operations may be left with fragile applications which can impact not only the users of that application but also other customers, maybe even bringing the system down with it. Even if this is not the case, unless they and the end-users have been involved in final acceptance testing and have applied all conceivable permutations of circumstances to the applications, as well as performing volume testing, the system may fail when it is actually in use. According to recent surveys, over 60% of systems fail to deliver what they were designed to do.

Service Level Agreements are discussed within a development context in Chapter 3: but quality - especially in terms of availability and performance requirements - should be built into the specification and should be part of the acceptance testing.

The same is true of modifications to existing applications or upgrading to new versions of application software. A formal change procedure needs to be applied and acceptance of changes, of the modification or the new version should include service level aspects.

In concept, operating systems and networks system software should be treated no differently. All new systems software, modifications or new versions should be subject to change control procedure and acceptance testing.

Change control procedure should be applied not merely to software but also to hardware, documentation and even to procedural, organizational and personnel changes as well as to changes of key suppliers (especially maintenance). Incorrect or ambiguous documentation can cause unavailability just as easily as hardware or software failure can. This applies particularly to documentation of Operating Procedures. Similarly organizational change could leave key functions uncovered and personnel change could lead to the loss of key areas of expertise. Establishing new contracts with suppliers and changing suppliers could result in service outage from unavailability of material or essential services.

2.3 Problem Management

Problems may hit availability or response. Unless the problem is properly managed the period of unavailability or poor response can be protracted . indeed, it can almost become a way of life and may eventually become grudgingly tolerated by the customer. There therefore needs to be clear procedures for reporting problems to a central point within Information Technology and Communications (e.g., Help Desk[3]); for these problems to be logged, prioritised and pursued through to resolution with appropriate escalation procedures. It is important to note that all problems should be logged, even trivial ones. The easily resolved, apparently trivial problems may indicate an undercurrent of customer irritation with the service that can readily be addressed by a simple solution such as additional training or better documentation.
Downtime means that something has gone wrong: in order to improve availability, we need to conduct a thorough problem audit of installation, learning from things that have gone wrong in the past identifying vulnerable targets.

If a database of problems is maintained, searches can be made problems relating to particular items of hardware or software or spec services, and this will aid problem audit.

2.4 Critical Component Failure Analysis

One method of conducting this audit is through Critical Component Failure Analysis.

A critical component can be defined as:

---

[3] Hiles A.N. and Gunn, Dr. Y. *Creating a Customer-Focused Help Desk: How to Win and Keep Your Customers.* Published by Rothstein Associates Inc. 2000 ISBN 0-9641648-6-8.

> "A path, piece of equipment, item of software or material; knowledge; a supplied service or a person without which a computing service would become unavailable."

During SLA negotiations with customers, availability targets will defined and the need for improved availability management is likely emerge. Processing paths required by the customers will need to be identified and the criticality of components in that path will need to firstly determined, and subsequently monitored. To improve availability these critical components need to be identified and resilience put in place.

Availability of **all** components in the chain needs to be considered, so improving the resilience of the most obvious critical components may simply reveal another criticality further downstream. Critical Component Failure Analysis needs to be undertaken of:

- firewalls
- local terminals
- data- and telecommunications equipment
- network bridges and routers
- network routing
- network software
- controllers
- distributed processors
- front end processors
- I/O devices
- mainframes, mid range equipment
- servers and desktop equipment
- plant
- operating system software
- applications programs
- individual databases
- support staff
- procedures and documentation
- supplied services
- personnel.

Typically, key weak spots may be:

- bandwidth or capacity limitations
- inadequate disk capacity or I/O bottleneck
- interruptions, spikes or dips to power supplies
- where water chilling is required, single supply of water
- inadequate air conditioning capacity
- reliance on a single printer, plotter or other key peripheral
- lack of alternate telecommunications routing
- undocumented procedures, processes or systems
- lack of cover for key staff (i.e. lack of cross-training or headcount restraints).

For each area, it is necessary to establish the actual availability of the critical components and seek answers to fundamental questions:

- who "owns" the component (for example does the computing service own it or the customer)?
- who should "own" it (in the sense of accepting responsibility for its availability)?
- what is its Mean Time To Fail (MTBF)?
- how many times has it been down in the last one, three and six month periods?
- at what times?
- what is its percentage theoretical availability during:
  - prime
  - non-prime
  - weekend?
- is that the percentage availability to the customer of that component?
- if not, what was preventing the customer from using it during that theoretical availability (i.e., was there another critical component which failed downstream)?
- what did that cost the business?
- what is its Mean Time To Fix (MTTF)?
- are there any alternatives to that component?
- how can alternates or resilience be built in?
- at what cost?
- what would the absence of this component cost per hour/day/week in a worst case scenario?

The answers to these questions will enable customer and computing service to establish what (if any) spend is reasonable to improve availability to the service level preferred by the customer.

Availability management may be improved in some cases simply by keeping a spare (e.g. a spare compatible modem) or by improving the maintenance call-out arrangements.

Critical Component Failure Analysis will reveal the theoretical availability of the service. A simplified calculation is in Table 2.1 below.

Table 2.1: Critical Component Analysis - Cumulative Availability

Component	Component availability	Cumulative availability
Network: WAN & Internet	99.95%	99.95%
Infrastructure Data Center / Systems	99.95%	99.95%*99.95%
Application	99.95%	99.95%*99.95%*99.5%
Total Availability		**98.85%**

The above example may be simplistic for some: where client hardware, client operating system, client middleware, client application, database system, server operating system, server middleware and server applications are involved, the end-to-end availability can be diluted to less than 98%.

If each component fails at the same time and is fixed in parallel, the availability will differ from a situation where each component fails sequentially or at different times.

Availability may therefore be defined in terms of consecutive hours of outage (e.g., a SLA target of failure less than 2 consecutive hours will be met regardless of whether there is more than one failure of up to two hours). Customers, however, may wish to define the SLA in terms of cumulative hours of outage.

The more sophisticated service vendors may use a mathematical modelling tool, Monte Carlo Analysis, that will indicate the probability of two random simultaneous failures and will factor this into their service levels. Table 2.2 provides contacts for Monte Carlo Analysis tools.

Table 2.2: Contacts for Monte Carlo Analysis Tools

1. Simtools is freeware add-in for Microsoft Excel. It adds statistical functions and procedures for doing Monte Carlo simulation and risk analysis in spreadsheets.
   http://www.kellogg.nwu.edu/faculty/myerson/ftp/addins.htm

2. Palisade @Risk provides risk analysis, decision analysis, optimization software, and other functions
   http://www.palisade.com/html/cd_request.html

3. TreeAge Software, Inc., publishes DATA 3.5 and DATA Interactive decision analysis software.
   http://www.treeage.com/

4. DACI Designmaster is a multi-tool risk analysis package generating worst-case scenarios, probability plots, sensitivities, optimum values and graphs.
   http://www.cyberspy.com/~daci/

5. Accurisk offers interactive high dimensional online modeling, numerics, monte carlo simulation, valuation and risk analysis of financial securities and portfolios
   http://www.accurisk.com/

6. Pertmaster provides project management, planning and schedule risk analysis software.
   http://www.pertmaster.com/index.htm

7. Clockwork Solutions provides predictive modelling technology and services
   http://www.clockwork-solutions.com/

2.5 Relationship with Security and Contingency Planning

At this stage the lines between Availability Management, Security, Risk Management and Contingency Planning become blurred.

Good IT and telecommunications managers have, in the past, been prudent enough to develop IT Disaster Recovery Plans involving alternative sites, communications and equipment to be used in the event of a disaster. However, the stakes have moved on. Now Disaster Recovery is too later for many organizations – the availability service levels and potential for loss mean that the organization simply does not have time to recovery: it cannot afford a disaster. In these cases, the goal has to be continuous availability. Equally, recovering the technological infrastructure without the business itself may not help corporate survival, so the concept of Disaster Recovery has been extended to cover all mission-critical business activities and infrastructure. Disaster Recovery has been incorporated into Business Continuity Management[4] and increasingly Business Continuity Management is being incorporated into a holistic approach to corporate risk, which we have called *Enterprise Risk Management*.

When implementing SLAs it is important to consider the impact of their breach and to create escalation processes from service failure to Disaster Recovery or Business Continuity plans.

In establishing the appropriate service levels for a SLA, it may be necessary to modify existing security or contingency planning arrangements. It may equally be necessary to reduce some of the present risks that the installation faces.

---

[4] Hiles A.N. ***Business Continuity: Best Practice.*** Published by Rothstein Associates Inc. 2000, ISBN 0-9641648-3-3 updated 2001. This book explicitly covers all the ten areas of business continuity competence required for membership of the Disaster Recovery Institute International (DRII) and the Business Continuity Institute (BCI) and is endorsed by both bodies.

One step on from the critical component failure analysis is the approach taken by the Swedish Vulnerability Board as described by the Swedish Federation of Data Processing Users, Stockholm. Their "security by analysis" risk methodologies can equally be applied to maintaining availability targets.

The Swedish Vulnerability Board's approach is broken down into a number of separate areas.

Start

This methodology is one of identifying risks. These risks may be external or internal. Such a risk analysis is a pre-requisite for a disaster recovery and contingency plan and should in any event be undertaken by an installation. It may reveal areas where risk can be reduced and hence the probability of availability increased.

Dependence

This stage examines corporate reliance in the systems and the extent of the organization's exposure in the event of their failure.

The Auditor

A comprehensive guide for auditors is part of the approach and auditors are encouraged to look at the security and continuity aspects of the systems.

Legislation

The final area provides a framework whereby existing and new legislation is summarised and system and procedures are compared with the requirements of that legislation to assess whether or not they conform to it.

Whatever the method, a Security, Risk Management and Contingency planning review is recommended before embarking on SLAs. The present position has to be identified so that arrangements can be compared to customer's requirements which will be established as part of the SLA negotiations. Quality service can only be delivered if security management disciplines are in place.

2.6 Scope of Service

What services are the SLAs going to cover? What will be the hours of service? A full SLA will cover:

- online services
- batch services
- network services
- distributed processing and equipment
- output arrangements
- help desk and technical support
- security services (e.g., virus protection, central backup of customer's PCs and workstations; media vaulting etc.)
- special requirements
- personal computing
- customer computing
- Internet / Intranet based services
- training.

Consideration needs to be given to the ultimate structure of the SLA. If the SLA is between IT and an external IT service supplier, it may be written in technical terms referring to each key infrastructure, platform, network or protocol (what we might call "service verticals"). Where it is between a technical service provider and a business user (e.g., Vice President Finance) then it is more meaningful for the end-user to have the SLA written in business terms. The Finance Service, for instance, might use a General Ledger package on a mainframe, sales and payroll input from different servers with important spreadsheets and reports using a local PC and printer over both Wide and Local Area Networks. The Vice President Finance is not concerned with the platforms the service uses – only in the end-to-end service. The SLA with the Vice President Finance might, therefore, focus on the business deliverables of the service (what we might call "business horizontals" that go across different platforms, applications, etc.) His or her SLA might be along the lines:

- payroll output will be transferred to the bank by 15.00 hours on the last Friday of each calendar month without exception
- invoice payment run will be made every Wednesday with output transferred to the bank by 15.00 hours each Wednesday. There shall be no more than one failure of up to 12 hours in any period of three calendar months
- All IT Finance Services will be available 100% between the hours of 08.00 and 20.00 each on the last five days before each calendar month end and the last three weeks before financial year end

... and so on. Of course, terms like "IT Finance Services" will have been explicitly defined.

2.7 Service Products

If we think of a service, it can almost always be broken into a number of individual sub-services or service products, each with its own deliverable and potentially its own service levels associated with it. First, we can establish a generic category (e.g., Maintenance or Desktop Support). Desktop Support can provide an example. A Desktop Support Service could offer the following service products:

- Service Desk
  - Request Management
  - Problem Management
  - Access Services
  - Password re-sets
  - Service Accounting information
  - Advice

  Equipment
  - evaluation
  - acquisition
  - testing

  - configuration
  - installation
  - moves
  - projects (roll-outs)
  - asset management
  - configuration management
  - de-installation
  - disposal
- Software
  - acquisition
  - evaluation
  - acquisition
  - testing
  - installation
  - projects (roll-outs)
  - version control / configuration management
  - de-installation
  - disposal

… and a host of other services and sub services. Each of these could have service levels associated with them. The benefit of this approach is that it modularises service delivery.

We can create a Catalog of Service, with the generic headings of the overall type of service and then identify the service products that make up that generic category (as we have done for Desktop Support).

Just as bricks may be used to build a factory, a school, an office or a home, so these modules can be put together to tailor the service to the needs of individual customers. It might not always be necessary to use each one of the service levels in a SLA with the customer but, knowing what they are, the supplier can develop a coarser SLA confident that the component service levels will permit achievement of that coarser level. In short, this approach permits granularity of service levels.

2.8 Service Hours

The SLA needs to specify the hours when service will be available. Allowance will need to be made for scheduled downtime. Scheduled downtime may include:

- preventative maintenance
  - hardware
  - plant
  - software
- hardware change
- telecommunications change
- system software testing and installation
- environmental work, building maintenance etc.

The timing of scheduled outages can best be decided by surveying customer's critical periods and seeking to minimise impact on them. Whenever possible, scheduled outages should take place out of prime shift. This will have a cost in overtime rates for the work to be done.

The times of regular scheduled outages (e.g. preventative maintenance) should be published as part of the SLA, as on-line standing information and in service guides.

As much notice as possible should be given to any changes to service hours . unless urgent work is required at least 48 hours notice for minor changes and perhaps a month or more for major changes.

With a growing tendency to automated operations and "darkroom" running, consideration needs to be given to the impact of failure during periods of unmanned service. Most disasters (and the most expensive ones) happen over the weekend. It may be necessary to invest in intelligent environmental monitoring systems . software usually running on a PC to which all warning and alarm systems are connected (*fire,* water and intruder detection systems, air conditioning, plant monitoring etc). Typically such a system would have an uninterruptible power supply and would take remedial action and invoke call-out arrangements, telephoning on-call staff or emergency services.

A differentiation may need to be made between service levels during manned operation and unmanned operation and between technical and customer support service levels . Help Desk, Information Center, etc. in prime and non-prime time.

Since groups of customers will have different working periods, cover may be required during an extended working day (perhaps from 0800 hours to 1900 hours) to cater for overtime working by the customer. Executive Information Systems, Decision Support Systems, office and electronic mail systems may be particularly demanding.

Service levels needs to reflect achievable targets and realistic allowances need to be made for problems arising during unmanned periods.

2.9 Real Time Interactive Services

The standard hours of availability for online systems need to be established. Availability targets should be set for these standard hours with, perhaps, a lesser target for services outside of those hours.

Response times for online services need to be set and the keynote should be consistency of response. It may however be advisable to set longer response times for peak periods and to explain to customers that response will vary according to machine loading. Workload limits can be set beyond which the response cannot be guaranteed.

It is increasingly common to set response times not for an individual transaction but in terms of a "basket" of transactions. This helps to smooth out the occasional aberration.

Online transactions can be broken down into different categories with individual targets - for instance, "90% of transaction type A to be completed within one second."

Particular care should be taken when setting targets for database systems: customers may make complex enquiries on a large database without appreciating the resources this uses and they may have unrealistic expectations of response time from the systems for such enquiries. Limitations on response time targets should be clearly established. Similarly, availability targets have to allow, where appropriate, for bulk update of a large database system or for the system to be frozen while backups take place.

2.10 Batch Services

Standard schedules for batch work need to be established. It should be made clear that processing schedules can only be kept if the customer adheres to agreed deadlines to submit input to the system. Equally, volumes of data need to be specified above which it may not be possible to maintain the schedule.

Accuracy requirements may also need to be stated: a high level of illegible or mis-keyed input could result in a substantial error or reject rate and consequently missed targets.

Arrangements for handover of data from customer to the service supplier should be specified together with authorization procedures. Run-to-run controls should also be identified.

2.11 Output Arrangements

There has been talk of the paperless office for well over a decade and most organizations are little nearer to it. However, negotiating an SLA does give the opportunity to question the need for hardcopy paper output as opposed to online information or microfiche. It also gives the opportunity to establish whether output is:

- timely
- relevant
- accurate
- in the format most convenient to the customer
- cost effective.

Many installations produce reports simply because they have been doing so, unquestioned, over the years. We have all come across instances where these reports are filed, sight unseen. Yet more frequent are examples of individuals keying data from these reports into a PC to massage with a spreadsheet or to produce graphics from them. If business analysis skills are brought to bear during the SLA negotiation it might be possible to provide exactly what the customer needs more productively.

Schedules will be required for production of output, together with instructions for handling sensitive or security classified output. Quality control of output especially of graphical or plotted output may need to be included in the SLA. Escalation arrangements should be made to cover the timescale by which the customer should notify the computing service of late output.

Delivery methods (courier, post, etc.) should also be agreed for the SLA.

2.12 Telecommunication and Network Services

Network services may be individually identified, with separate service levels applying to each service.

However, since most users are concerned with the response at their desks it is not uncommon to bundle in the network services with the application as part of a single SLA.

One of the main problems is monitoring availability and response through the whole of the network. This is particularly true of a remote customer accessing the computing service through intermediate computer installations. While it may be possible to monitor the availability and response at any particular component of the network, it may be easier just to measure the overall response at the customer end. The ownership of network devices downstream from the computing service needs to be clearly established and responsibilities for invoking maintenance arrangements have to be agreed. Given a choice, it might be advisable for the computing service to accept 'A' end responsibility for telecommunications (i.e., the end authorised to invoke call-out) for services owned by the customer. Accepting 'A' end responsibility (i.e. the end authorised to give instructions to the telecommunications service vendor) doesn't prevent the bill for telecommunications services being sent direct to the customer!

Figure 2.2 identifies definitions for Telecommunication SLAs. Transit Time can be measured across Customer Premises Equipment, to router, across the network "cloud", to the destination router and hence to and across the recipient Customer Premises Equipment.

Figure 2.2:

Definitions for Telecommunication Service Levels

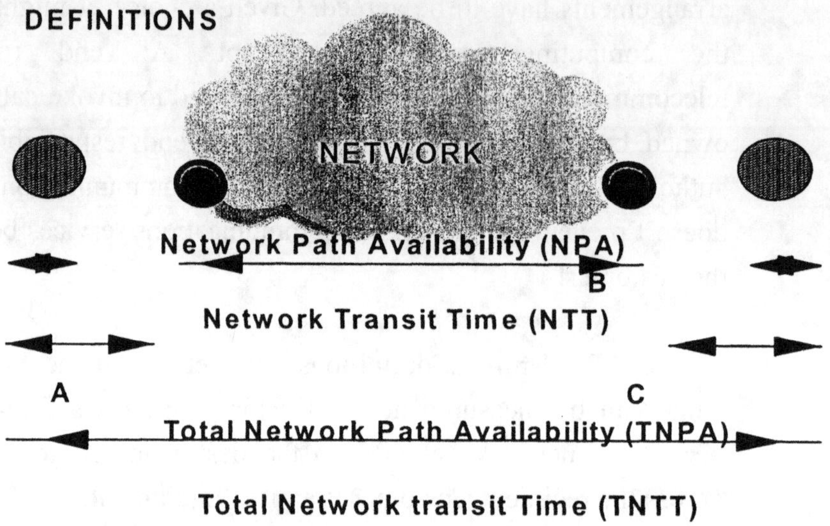

TNTT = A + B + C

SERIALISATION DELAYS - Depend on access speed

ROUTER PROCESSING DELAYS - 30-50 mSec

Options for availability include:

1. Host, single line & router, ISDN backup & remote site with no backup
2. Host, single line & router, ISDN backup & remote site with ISDN/PSTN backup
3. Host, dual lines, dual routers, remote site no backup
4. Host, dual lines, dual routers, remote site with ISDN/PSTN backup

5. Host, dual lines, dual routers, remote site with dual lines, dual routers.

These options achieve differing availability. As a guide:

ISDN backup, tail circuit:	99.3%
Single router:	99.46%
Dual routers:	99.86%
Total network:	99.850%
Site to Vendor Serving Office:	99.900%
Site to Vendor with dial backup:	99.995%
Frame relay cloud:	99.998%
Vendor's Premise Equipment:	99.990%

Other service measures include:

- latency (the Network Transit Time for a 128-character-length package)
- transfer time for a file of defined size (e.g. 5Mb)
- other measurable data includes:
  - Committed Information Rate (CIR)
  - Committed Burst (Be)
  - Bandwidth Interval (Tc)
  - Excess Information Rate (EIR)
  - EIR = Be/Tc

Response may be defined as "the time elapsed between a user device sending a message and its receipt." Typically host and database processing time is excluded from response measurement.

2.13 Outsourcing

Customers are currently responsible for well over 50% of IT expenditures. Increasingly they have the discretion to buy packages, turnkey systems stand-alone processors, application services and outsourcing services.

The in-house Information Technology service has an opportunity to provide facilities management services for equipment and applications. If it does, it may be in competition with external outsourcing contractors.

In this event, a SLA will be required to define service standards.

For organizations unable, or unwilling, to invest their own capital in the communications, equipment and applications necessary to run these applications, outsourcing is an obvious solution. For most of them, there is little Return on Investment on Information Technology: removing capital spend from their accounts makes the balance sheet look good to financial experts and can enhance the organizations share value and avoid or reduce the need to raise expensive capital on the financial markets.

So outsourcing can make sense – especially since it also avoids the need to hire, recruit and train staff or, where partial outsourcing is considered, it may greatly reduce the complexity of a customer's IT infrastructure and operations.

However, in this case the supplier cannot be held responsible for failure or performance problems with areas it does not control – such as their customer's service infrastructure. This may include the customer's LAN, workstations and Web connection. It is therefore even more important than normal to have in place tools, processes and procedures that enable quick diagnosis of where the problem lies – and effective SLAs to ensure that, where they lie with the supplier, they are promptly rectified. So reporting on service non-compliance needs to be backed up by prompt and effective issue escalation and resolution.

For problems arising from lack of capacity, providing more memory or disk space may be a solution: but there are cost implications. These need to be considered in the contract or SLA with the supplier. What are the rules?

Connectivity problems may wreck a service, especially in online mall and similar buying or trading applications.

Studies by Birbeck College and others indicate that over half of outsourcing contracts suffer serious dispute between supplier and customer and 70% of IT outsourcing users hit problems. Half of the disputes are resolved by changing vendor; 22% are terminated by customers bringing the service back in house.

Recently industry watcher Sage Research, based in Natick, Mass., published a survey that said that disputes between customers and their service providers involved services like Web-hosting, VPNs, and e-mail. The reliance of these services on the Internet shows just how important it is to get these aspects of the SLA right.

The cause of outsourcing problems lies in a few key areas:

- Sometimes the solution to a problem area is seen as outsourcing. Often cost data is inadequate to put a real cost on each element of the service and the service quality is not adequately measured. In this case, when the service provider has cleaned up the operation, it will be reaping high rewards, while providing mediocre service.
- Sometimes the main driving forces for outsourcing are political and cash saving: if outsourcing does not have strategic and operational logic as well, the real benefits will not be achieved. Unexpected charges are often a cause of contention and these often stem from inadequate service definition in the first place. Emphasis on cost saving forces cut-throat competition and, all too frequently profit has to be sought from anything which has not been clearly specified (known to the cynical as "value added sales"). This can mean that the expected cash savings are made on the prime contract, but anything not clearly specified is a chargeable extra. In one case, the (unspecified) ad hoc payroll reports cost more than running the payroll!
- In our experience, many contracts (even in those valued in excess of $50m) details of the service to be provided and of service quality expected can be woefully inadequate or ambiguous. Expected service levels are not achieved; contracts are sloppy. Sometimes this is a result of drafting these documents without reference to the existing specialist operational staff who really know the detail; but, even if existing technical staff are consulted, they may not be sufficiently expert in drawing up specifications and service levels. Particularly in the public sector service culture, historically the emphasis may have been on service provision rather than on service measurement The proliferation of customer charters has led to concentration on quantification at macro level: but to manage service, detail is often required at much lower levels.

- Changing requirements, sometimes not able to be anticipated at the time of contract, but often able to be mitigated if the contract had been better thought out.

So, how can we make sure an outsourcing deal does not hit the rocks?

By benchmarking the in-house service, it is possible to get a reasonable idea of its competitiveness before going out to tender. If the service is a problem area, it is usually cheaper to get it right before outsourcing: otherwise cost savings and quality improvements will (rightly) be credited to the outsourcer. It sometimes happens that a conscious decision is taken to outsource for a limited time, have the outsourcer re-organize and improve the service and then to bring it back in house: if this is the objective, the contract duration needs careful consideration. A short contract is likely to carry a heavier annual price tag than say a three or five year contract.

By focusing on mission achievement and the critical success factors that determine this, we can more clearly identify priorities and strategies for outsourcing. All scenarios of service provision can be considered and all options evaluated - including in-house options. We can ensure that the customer maintains control of the strategy. Giving strategic direction of a service to an outsourcer can be a little like appointing a chocaholic as manager of the candy store! The risks associated with outsourcing, as well as the benefits, need to be carefully evaluated. If outsourcing is the solution, what is the basis: contractor, partnership or something in between? It is the experience of many customers that so called partnership arrangements are often insufficiently explicit and leave too much ambiguity, raising expectations on both sides which cannot be fulfilled. That is why many customers are moving from partnership to more of a contractor relationship the second time around. Tight service specifications and service levels do not necessarily prevent the supplier from being proactive and creative.

We can tighten up the contract, the service specification and the service level agreement. Sometimes Service Specifications, Invitations to Tender, Contracts and Service Level Agreements are drafted in isolation from each other and by different people: if these activities are fully coordinated, we can ensure all documentation is complementary, consistent, comprehensive and unambiguous. Often the Service level Agreement is drawn up after the contract is placed - by which time it is too late. If it forms an integral part of the contract (preferably as an appendix) then expectations of both parties are clearly set. If the outsourcer is allowed flexibility to deliver the result rather than prescriptively describing the process, they can be creative, flexible - and maybe even make a reasonable profit without prejudicing service quality. Unless there are legislative, regulatory, health, safety or environmental issues we should not necessarily be concerned about how the results are achieved, as long as they are.

If the contract identifies a charge for every service provided by the supplier (whether apparently included in the service specification or not) then there should be no hidden surprises in the event that additional services are required. The contract term (which should rarely be more than 5 years) should reflect the stability and anticipated life span of the customer's activities and frequent review of contract and service levels should be built in. It is most unusual for a contract or service level agreement to remain valid for the entire duration of a contract so change should be anticipated.

Knowledge of the outsourcing market is essential, both to establish the suppliers to invite to tender and for evaluation. If supplier evaluation criteria are solely or mainly dominated by price, supplier selection may be flawed. Other considerations need to be weighed, including vision, technical competence and capability, financial stability, knowledge of the sector, compatibility of culture and motivation and record on staff issues.

Where a service is already provided in-house, the transition from in-house to outsourcing demands meticulous planning. People issues require tact, persuasion and effective communication. Staff need to be kept informed, especially in respect of implications on pay, pensions and security. If key staff leave, the transition may lurch. Procedures may need amplification. Problems will inevitably arise (both in transition and during the course of the contract). Both customer and supplier therefore need to implement and exercise excellent issue escalation and management procedures.

Too often, there is insufficient skilled resource retained in-house. In a medium sized contract, the supplier can often have half a dozen people working on account management and customer relationships. What makes the customer think they can manage with one contract manager? The service levels achieved need to be carefully and regularly monitored and it takes resource to do this. The relationship with the service provider needs to be worked at throughout the contract - maintaining alignment of objectives and relevance of service, cultivating openness and understanding. Market knowledge needs to be maintained through the life of the contract: knowing that the customer is aware of the competition, the supplier remains alert and responsive.

Outsourcing represents no panacea for cost saving: if undertaken with care and for the right reasons, it is, however, a useful option for management in mission achievement.

The lesson from experience is that there is a high probability of dispute arising in any outsourcing supply – which points up the case for including effective mediation or arbitration clauses in the contract.

2.14 Applications Development Services

Since special circumstances surround service levels for applications development services, a separate chapter (Chapter 3) is devoted to this area.

2.15 Distributed Processing

The computing service may have responsibilities for supply, installation and maintenance of distributed processors, communications devices, printers and terminals.

It is notoriously difficult to maintain an accurate asset register of distributed equipment but this is nevertheless essential. Responsibilities for remote hardware have to be clearly defined both for the equipment and for any consumables and supplies. The asset register could be extended to incorporate maintenance arrangement and call-out procedures (and it could then be incorporated into the Disaster Recovery Plan).

2.16 Help Desk and Technical Support

The Help Desk should be the first point of contact for any customer with a problem. A Help Desk may be little more than a receptionist function receiving calls and diverting them to the appropriate specialist. On the other hand it may be fully competent to resolve the majority of problems which it receives. It is highly unlikely, however, that the Help Desk will answer all the queries it receives and inevitably some problems will be referred to specialists within the computing service or possibly to external supplier Problem Management Procedures should be in place to ensure that "ownership" of a problem is registered and that no problem can be un-owned". "Ownership" of a problem may either reside with the Help Desk regardless of where it is referred or "ownership" may pass to the person to whom it is referred. In either case a Problem Manager should be appointed to review outstanding problems and to escalate them when necessary. The customer will be a party to deciding how to prioritise problems on the applications or services they use. The customer may also help to set escalation procedure.

2.17 Internet and Intranet Based Services

This section covers SLAs relevant to in-company Intranets; between-company Extranets; Internet Service Providers (ISPs); Application Service Providers (ASPs); Managed Service Providers (MSPs); and Total Service Providers (TSPs). They are covered in more detail elsewhere.[5]

Internet based services, whether ISPs, ASPs or dot coms play in a difficult and risky market.

---

[5] Hiles, A.N., ***E-business Service Level Agreements: SLAs for ISPs, Strategies for Service Providers, E-Commerce and Outsourcing;*** Rothstein Associates Inc., 2001, ISBN #0-9641648-9-2

Time to market is deemed crucial – yet the old saying "you can have it cheap, you can have it right, or you can have it now" remains true: at best, usually only two of those three are attainable. At worst, two are sacrificed for speed to market. The Application Service Providers are trying hard to prove all three can be achieved. There is danger, too, in quick time to market – it is a matter of fact that most of those first in the market do not succeed: the hare is overtaken by the tortoise who sees, and avoids, the pioneer's mistakes.

While growth of voice traffic is growing at 10% a year, IP traffic is growing at 1,000% a year. Individual growth of a service or a supplier can be unpredictable and dramatic.

There are many inter-relationships and inter-dependencies. To the customer, this spells danger. While SLAs cannot by themselves ensure continuity of supplier, SLAs are imperative with and between e-service suppliers to provide a degree of security and consistency of service quality.

E-business is growing and becoming increasingly an indispensable part of business activity. Loose virtual relationships are growing alongside more formal outsourcing and contract-based services. What is still missing, in many cases, are meaningful Service Level Agreements (SLAs) that are unambiguous, comprehensive, and enforceable.

Typically SLAs will cover:

- Availability timings and service levels
- Maintenance and scheduled service outages
- Reliability (i.e. number of incidents of outage)
- Packet loss
- Latency service levels
- Outage resolution times (possibly in different priorities)
- Escalation procedures (possibly in different priorities)
- Customer / user satisfaction
- Service Performance Review items.

Key weaknesses in existing Service Level Agreements include:

- Specification of effort or targets rather than of results
- Unclear service specifications
- Unclear definition of service level metrics
- Inappropriate or misleading measurements.

2.18 Security Services

A menu of services offered should be agreed with customers. These may include:

- provision and management of firewalls
- virus protection and clean-up
- password management
- security reviews of LANs and PC installations
- management of file servers
- advice on access control standards and techniques for customer owned applications
- facilities to backup PC and workstations to a mainframe or distributed processors
- backup of distributed processors
- storage of magnetic media (media vaulting)
- backup of user owned databases to an alternative processing site
- disaster recovery and contingency plans
- advice on legislation (e.g. security requirements of data protection or compliance with security requirements of financial legislation).

2.19 Special Requirements

Where dedicated equipment or software has been acquired for the customer or where non-standard support is provided, the Service Level requirement needs to be included in the SLA.

2.20 Personal Computing

Personal computing services are frequently a bone of contention between the computing service and the customer. Response on office systems can vary widely depending on the load on the systems. Where mainframe spreadsheet, personal database facilities, and word processing are provided they are frequently unfavourably compared to the accessibility, availability and ease of use of PC tools.

Support of PCs is an equally thorny issue. In many organizations there is no strong central control of PC purchase and PCs can be bought on departmental budgets. These PCs may be connected onto a LAN also bought on the departmental budget and the computing service may then have the problem of connecting individual PCs or the LAN to the mainframe service. It is increasingly becoming the norm for terminals to be PCs: non-standard PCs can take a disproportionate amount of support.

This emphasises the benefit of standardizing on PC equipment and LAN architecture. If the computing service has control of PC purchases it can set the standards: otherwise it can only be advisory. The SLA should aim to encourage the use of standard PCs and of standard PC software in order to reduce the range of equipment and software that it is called upon to support.

It should also be clearly stated who is responsible for backup of PC based systems: usually this lies with the customer.

Over the years this Author has asked delegates at Security Seminars and Conferences how many have been affected by virus: the figure at present is running at over 70%. Viruses are a real problem and rules need to be laid down to ensure that customers do not infect other customers' PCs, do not impact other customers on the same LAN or impact the computing service. Services which are being offered include:

- tight central control through network management tools which limit or block user's ability to effect changes ("back to the glass house")
- automated equipment and software audits
- remote booting
- diskless PCs

- enforced or optional use of server disks rather than using the PC's own hard disk drive
- virus checking services
- virus cure services
- mainframe backup.

Where personal computing support is offered by the computing service, the opportunity can be taken to apply conditions that may include:

- banning games software
- banning public domain software
- banning package copying except for backup
- specifying the packages and hardware to be supported.

2. 21 Customer Self Computing

With the growth of fourth and fifth generation languages, easy-to-use report generators and PC or server based relational databases and spreadsheets, many customers are writing their own applications. Acceptance of responsibility for running these applications could be made dependent upon quality assurance of such systems. Formal production acceptance and change management procedures should be applied to such applications whenever practicable.

2.22 Training

The customer requirements for training and the computing service's terms a conditions of providing it should be included in the SLA.

CHECKLIST #2.1
SERVICE LEVEL QUANTIFICATION

1. Quantify customer satisfaction with:
   - Access
   - Availability
   - Response
   - Change management
   - Support
   - Problem-solving
   - Services/products
   - Cost.

2. Support Functions Quantify:
   - "Noise level"
   - Number of problems
     - resolved within target
     - unresolved
   - Frequency of user meetings
   - Computing Service products and services (see also 5)
     - identify all services
     - identify cost of each
     - identify value of each
     - identify usage of each.

3

HOW SERVICE LEVEL AGREEMENTS APPLY IN AN APPLICATIONS DEVELOPMENT ENVIRONMENT

3.1 Applications Development

SLAs are normally thought of as applying to an ongoing production system. Production SLAs can be made more effective and service provided cost-effectively to a higher quality if the SLA concept is embraced during systems development.

Today there is growing pressure for development to be oriented to business solutions and to be delivered quickly and cheaply. Since a good systems analyst will only capture some 70% of the specification requirement, there is also a pressure to prototype (which can capture over 90% of the requirement). Over *50%* of computing spend is in the hands of customers of the computing service, so increasingly developments are customer driven. They are subject to moving targets depending on rapidly changing markets and changing legislation. Speedy development is seen as requiring 4GL and CASE tools, relational databases, and a reduced emphasis on formal ("straitjacket") standards. To reduce downside risk and to avoid runaway projects, there is pressure to implement in stages.

These characteristics tend to militate against service level criteria being specified at the design stages of a development project and checked during development.

But availability, usability, throughput and other service level issues are crucial to the success of any development project. The price for not doing so may be high:

- according to Business Week, "Nearly half of new software is so badly flawed that the company is not able to do what the program was supposed to support"

- John Spiers, Product Manager of database company Oracle, claimed "Some users have bought the wrong database system and had to buy another, on the basis of incomplete evaluation or promises not fulfilled by a supplier. In some situations people have lost millions"

- a report for the British Government's Department of Trade and Industry claimed UK organizations wasted £500,000,000 a year by buying the wrong hardware and software.

Poor service level achievement may not be the only reason for these failures but it is a significant contributing factor in many of them. How can we avoid similar mistakes?

Quality is the key. If quality is built into systems from feasibility stages on through to implementation, not only will those systems be more effective, they will be more robust and more serviceable.

Applications bugs are a prime cause of applications downtime . even if they do not take the application out, they can make it unusable or impossible to support to a given Service Level. Various surveys show that it can cost up to one thousand times more to correct a bug in a production system than it costs to eliminate it at design stage.

Other surveys show that even a good systems analyst is likely to capture only 70% of the requirement. The inevitable communication gap between analyst and customer contributes to this. Prototyping is claimed to improve requirement capture to over 90%, but it is axiomatic that the customer only really knows what he wants when the system has been delivered and he can experience its shortcomings. Over 80% of bugs are caused by inadequate specification and over 80% of bugs are resolved by changing the specification or design . that is, by putting up with them!

Quality has a payback: after implementing quality methods, IBM saved over $2,000,000,000 in a 3-year period . big numbers, but considering that 70% of software expenditures are said to be in maintenance, even an in-house development shop could show useful savings.

We have seen earlier that Information Technology, Infrastructure Management or Installation Management methodologies have to be in place before a SLA can be applied to production systems. In exactly the same way, Applications Development Methodologies have to be in place before we can commit to a SLA development environment . and, realistically, before we can deliver a given level of service to any specific application. Service requirements should be an integral part of the specification and, when software is bought in, acceptance of it should depend on meeting service level criteria.

We have to find a ruler in order to measure service levels of the development process. One ruler could be Function Point Analysis (FPA). FPA, devised by IBM, measures a number of "things" in the proposed development. It:

- counts the number of inputs, enquiries, outputs, master files, etc.;

- weights the sum; and,
- counts and adjusts for the complexity of project.

The result is a relative measure of function value and an estimate of number of lines of code a competent programmer will have to write to develop the application. Knowing this, and the number of programmers available and their cost, an estimate can be made of development timescale and cost. FPA enables SLAs to be structured to reflect price and maintenance costs, delivery timescale, bug rates and other aspects in terms of so many per thousand function points. Figure 3.1 illustrates the point.

Figure 3.1: Development using FPA

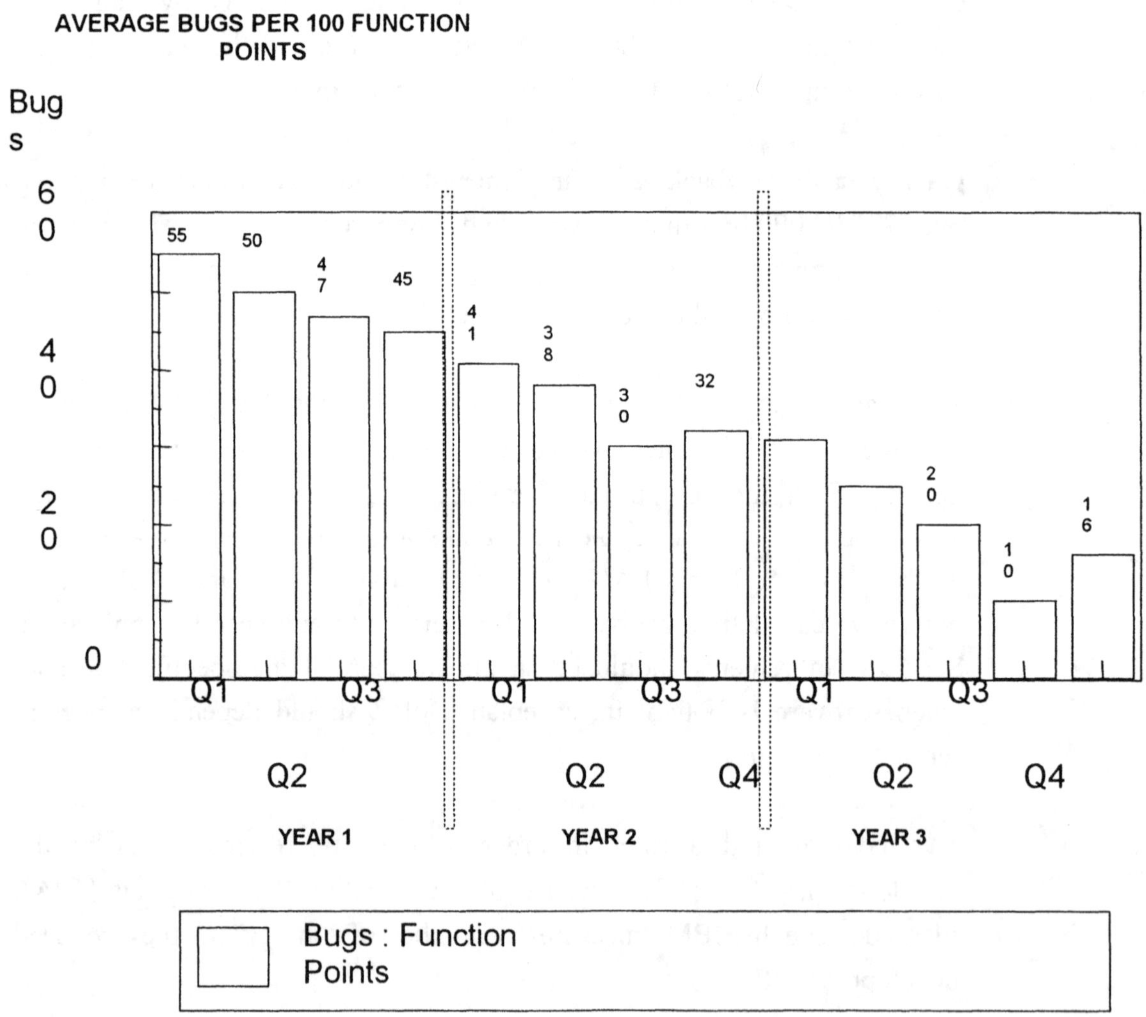

Other key approaches required in development follow:

3.2 Development Environment

The development environment should be one in which a framework overall corporate IT standards apply and, in particular, overall service standards are set with which developers are motivated to comply.

3.3 Feasibility Study

The Feasibility Study needs to assist in the definition of basic service criteria relevant to the proposed new system. These will be defined as development progresses. It should involve:

- establishing basic definition of application- specific service levels
- establishing downside risks of system loss
- considering performance issues:
    - response
    - turnaround
    - applications efficiency (in terms of capacity requirement)
- for database systems, data freshness requirements considering support requirements
- budgeting for ongoing support costs
- budgeting for project management costs
- budgeting for Quality Assurance/Quality Control
- budgeting for security
- including risk analysis
- assessing value of application
- assessing impact of loss of application once live
- considering total life of system relative to hardware life and possible

portability requirements.

3.4 System Analysis/Specification

The system analysis and specification stage needs to include:

- structured analysis
- establishing reliability service level (defining frequency/nature of errors permissible)
- establishing availability service level
- establishing throughput service level
- for database systems, specify data freshness (otherwise data currency . how up-to-date is the data)
- specifying minimum application life
- setting longevity requirements (e.g. by specifying that, even if software is bought as a package, written by a 4/5GL or CASE tool, final code should be in a Systems Application Architecture supported language)
- setting response service level
- accurate sizing and capacity planning to provide a robust growth plan in line (or over) forecast growth over the life of the application
- quality control
- quality assurance
- security - logical and physical access controls
- fault tolerance.

3.5 System Design

The systems design stage needs to include:

- structured design
- design for usability
- design for application longevity
- quality assurance

- compliance with service levels specified in terms of:
    - availability
    - throughput
    - response
    - turnaround
    - longevity
    - usability
    - sizing/capacity
    - portability
    - security
- quality control
- accurate and comprehensive documentation.

3.6 Invitation to Tender/Contract

The equipment and/or software evaluation, acquisition processes and contractual stages should reflect the service level requirements previously defined. In particular, they should:

- include service level requirements as part of the Invitation to Tender
- include service level requirements as part of the contractual requirements
- provide for a clause in the contract for retention of a significant part of the payment after delivery until acceptance testing (including performance to service level standards) has been completed
- include penalty clauses for non-compliance with service level requirements
- specify training requirement (IT and customers)
- specify documentation requirements
- specify support requirements:
    - maintenance (cover and response)
    - help desk/other support
    - life for which system is required to be supported (if software, duration of support for that version)
- investigate financial status of supplier and ultimate vendor to be

assured of their viability over the life of the system

- for software, include contractual requirements for escrow arrangements to assist support of bought-in software in the event of collapse of the supplier by making source code available.

3.7 Implementation

At implementation stage, from a service level viewpoint the key activity is testing to ensure the delivered system meets service level requirements. Acceptance testing therefore should include:

- volume testing (peak volumes and forecast volumes)
- permutation testing (all possible permutations of activity)
- performance testing (to service level specifications)
- usability testing
  - by Operations
  - by Customer
- sign-off from development and acceptance into production by Operations
- sign-off and acceptance by Customer.

3.8 Post-Implementation Review

The post-implementation review should include:

- performance against service level specification
- current utilization against forecast sizing
- anticipated growth against original forecast growth
- current impact of loss of service against that identified at Feasibility Study and Design stages
- documentation in use compared with specification
- adequacy of computing service and customer training.

3.9 Service Orientation

Customers may be more likely to assume service level requirements than to specify them: it is the IT department's role to ensure that all requirements are covered in the specification and to ensure they are met before delivery of the project.

Figure 3.2: Development Environment

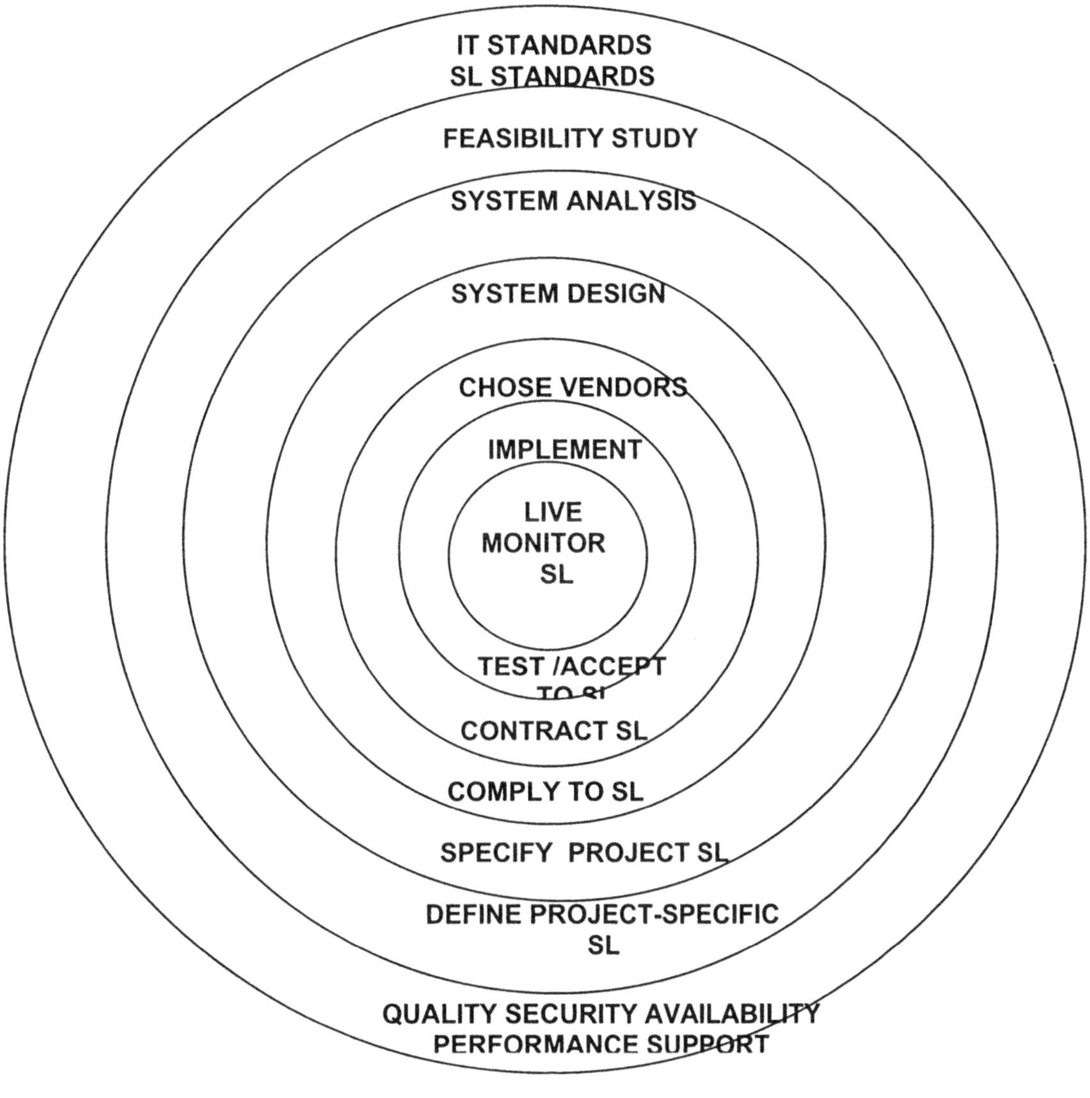

4

KEYS TO MEASURING AND MONITORING SERVICE; DESIGNING AND IMPLEMENTING AN SLA

4.1 Introduction to Service Measurement

The terms *Performance Measurement* and *Performance Monitoring* may sometimes be confused. Performance Measurement involves ascertaining the quality and reactive correction to meet service levels. Performance Monitoring is more effective, since it ascertains service quality in real time, although any correction to service quality is also reactive. *Service Management* is the most effective technique, since it both ascertains service quality in real time and takes corrective action in real time.

Performance Measurement, Monitoring and Management tools and techniques are essential to ensure service level compliance. Key areas to monitor are:

- Application
- Network
- Systems
- Customer satisfaction.

Measurement and Monitoring enable the enforcement of penalties and potentially provides evidence in the event of litigation. Performance Measurement and Monitoring and Management tools are identified and placed in context.

For simplicity we will refer to these tools in this Chapter simply as *Monitoring Tools*.

4.2 Measuring Performance and Availability

Service monitoring processes start with the hardware resource monitoring facilities. Most mainframes and mini computers, together with many peripheral devices, generate information about resource utilization and performance. In addition to these hardware monitors, system software may produce further performance information. The same applies to the network. An on-line application could perform some of its own response time monitoring by having the transaction time-stamped on its journey from the terminal, through the network nodes into the host and back to the terminal. This information could then be piggybacked onto the next transaction and extracted for analysis at the host.

Incident and Problem Management Procedures should also identify outages and their duration. The Help Desk may record problem information which also reflects downtime or system outage. Engineers' logs (both system-based and manual) contain yet more data on performance. The capacity planning function needs to establish utilization and performance in order to make sensible decisions about hardware sizing for the future: this data is again of value for service level management.

Performance and availability data may therefore be spread over a very wide area, involving different people perhaps from different department within the computing service. Bringing all this information together may not only make for more efficiency by avoiding duplication of effort but it can provide the basis of a service management database. There are a number of software tools available which access the raw material from system monitors and consolidate it into a database. By adding information from incident and problem reporting systems and from manual logs this database can be enhanced and can become the single focal point for system management.

From this Service Database we can draw historic information which will greatly assist the Critical Component Failure Analysis.

Since the Service Database may also contain utilization information it can underpin an equitable policy for charging out Information Technology services.

Trend information from the Service Database will help to identify area where availability and performance are declining: if regular trend report are produced the "hot spots" are quickly revealed.

Network monitoring is frequently more problematical. With a host-base4 network management system, extracting information may not always b easy. Host system methods can suffer a number of disadvantages:

- lack of standardization of downstream information
- inconsistency between different manufacturers in definition and presentation of faults
- difficulty (or impossibility) in identifying or cross referring faults reported simultaneously
- failure of the network management system if the host fails
- imposition by the monitor of a significant system overhead on the

host processor

- difficulty in host monitoring of downstream devices, the failure of which may impact service levels.

Current network monitors are available which may:

- be based on the mainframe or midrange systems
- be server based
- be PC based
- "heartbeat" to advise a host the equipment is available
- "ping" to establish remote equipment is available
- track the number of retransmissions arising from poor line quality, enabling rectification sometimes even before customers notice the problem
- show average response for devices being monitored, identified in various line groups
- monitor applications showing monthly response time trends
- monitor individual applications against the number of transactions
- show peak hour and average daily line customer utilization
- operate by echoing dummy transactions from the customer's terminal to identify "at-the-desk" response times.

4.3 Monitoring Tools and Their Use

SLAs are virtually useless unless actual performance can be measured, monitored and managed. Leading service providers undertake their own performance measurements and often publish them on their web sites. However, while such measurements are usually undertaken by the service provider, it is important that these are auditable or that they are checked and correlated with the customer's own measurements.

Performance monitoring tools use capacity so, in a heavily loaded service, they may impact response or latency. Equipment and network capacity should therefore be sized to take account of the performance monitoring overhead. This overhead maybe reduced by taking snapshots rather than running the performance monitors as background tasks. The longer the period between the samples, the greater is the dilution of performance data.

The following elements of a typical Internet service, for example, can be monitored:

- User / client
- Local Area Network
- Wide Area Network
- Web server
- Application server
- Database Operating System
- Server.

The customer is concerned with is end-to-end response and availability of applications. Availability may be monitored:

- Locally (at server level)
- Remotely, by logging in to the network and running a periodic artificial transaction (or, for response by running an artificial but representative bundle of transactions)

4.4 Application Monitoring

Application response has to be monitored to detect slow application response with its consequent impact on the customer. If this happens in online brokerage, credit checking or shopping the impact could be serious leading to potential losses.

Monitoring tools include Proxima Technology's sla*manage* and InteQ's InfraWatch, both using BMC's Patrol for e-Business. Another is Adlex NetworkDelta APM[SM] (Application Performance Monitoring) solution. This provides a real-time view of the customer's experience. NetworkDelta APM[SM] is a service management solution that provides, by application and user, visibility into both current and historical application performance. True measurement reveals the actual performance delivered to each of your customers. Diagnostics determine if application performance problems are due to internal issues or to poor service. Most companies that sell management tools and remote application management supply tools or services both for service providers and for customers.

These software tools can be used to identify performance issues and can contact support personnel by e-mail, mobile telephone or pager. They can also establish the root cause of the problem for use in causal analysis and elimination of these root causes (which could, for instance, be an excessively large index on a database).

Standard system and network performance monitoring tools may not be able to detect costly and disruptive application brownouts. VeriServ is an end-to-end, real time application measurement tool that does provide alerts to brownouts and end-user performance issues.

Use of applications management software enables SLAs to be founded on clear, objective, auditable measurements with the knowledge that compliance can be proved or disproved. It helps avoid downtime or poor performance and focuses both service provider and customer on actual service performance against contract.

4.5 Network Monitoring

Performance monitoring is now standard at the physical layer both for company's with their own extensive communication infrastructure and for telecommunication based service providers themselves. Providing back-to-back SLAs have been negotiated carriers can offer SLAs on service metrics, including latency, availability and reliability, across multi-vendor private networks SLA monitoring is an established practice at the physical layer. The core physical layer Internet services may be enhanced by value added managed services (e.g. Virtual Private Networks (VPN), managed router services or managed Internet services). Again, SLAs should be applied. However, an increasing number of ASPs now managing this sort of function in-house.

Network performance monitoring and reporting tools come in various shapes and sizes. Cisco and Nortel Networks routers have their own management tools and other vendors have products for monitoring performance of the Internet backbone. Automation of the collection, interpretation, and presentation of service level data can be achieved by using specialized tools such as InfoVista.

Firehunter/PRO from Agilent Technologies enables Internet service providers (ISPs) to measure the quality of their IP services precisely, assess these from the customer's standpoint, and manage them proactively. It keeps track of all relevant network and service data, and provides administrators with comprehensive, automated reports.

Another tool for remote access monitoring is Inverse Technology's Access Ramp code. Inverse is an independent company which has been rating ISP's and Internet access since 1996. Inverse collects data by compiling information on dial sessions connected to various cities through different ISP services. They also collect information using Access Ramp technology which collects data on dial sessions. Access Ramp is a portion of code that is included in the AT&T Global Network Dialer (v.4.17.1 and better). It measures and records information about a users dial session. When a user makes a successful connection to the network, this file is uploaded, transparently and transported to Inverse, who compiles it, and publishes the results. Inverse then rates the Internet Service Provider's (ISP's) based on five categories. These categories are:

- Percent of busies
- Ring no answers
- Modem problems
- Login failures
- Time to connection.

The information collected with Access Ramp may be compiled into reports that are provided to dial users at a user accessible web site to participants in some ISP's SLA programs. Different sets of reports may be available. There is a summary report that provides an overall view of all dial sessions logging into the AGN Network. Additionally, possibly at extra cost, a separate set of reports, which reflects specific information about an ISP client's set of dial users is provided. Access Ramp and AGN Network Dialer data can be used to provide custom reports, for instance, reports can be broken down by city, state, user, modem, type etc.

4.6 Case Study

Vigil offers a niche Internet based service in information provision. Its service automates the finding, gathering and timely delivery of strategic content for its customers, based on their individual, specific content criteria. Vigil's clients include Oracle, Bowing and Motorola.

The company hosts these systems at Exodus Communications' data center in Waltham, Mass. While Vigil was happy with Exodus' hosting services, it was also seeking performance management services. Exodus introduced InteQ. InteQ's InfraWatch™ was implemented in just one month to monitor, manage and report on performance for all of Vigil's hosted equipment.

InteQ's intelligent notification capability ensures the right person receives timely information on issues, even before performance is affected. Vigil claims that, for instance, the disk space monitoring function has enabled Vigil to prevent outage on more than one occasion. Vigil has a 98% utilization goal and they credit InteQ with avoiding optimising capacity use and avoiding hardware spend.

www.inteq.com

4.7 Systems Monitoring

Many tools are available to monitor systems environment in the data center. Such tools can be used to report on percentage uptime/downtime, number of outages, the longest and shortest outage, MTBF (Mean Time Between Failure), and MTFM or MTTR (Mean Time for Maintenance; Mean Time To Repair / Restore).

Tools come as default for various server environments (e.g. Citrix Resource Manager and Microsoft Systems Management – SMS – for Windows clients) and ASPs may use these.

Some server vendors offer their own tools for performance management and availability monitoring, for instance, Sun (Netmanager) and HP (Openview)

Some Application Service Providers (ASPs) are creating their own performance reporting tools incorporating products and, in some cases, the server default performance management tools.

Brixnet, a Chelmsford, Mass. based ASP, offers Brix System, which delivers end-to-end monitoring of customers' applications, simulating end-users' transactions at predefined intervals and keeping records of response time and connectivity issues.

4.8 Satisfaction Monitoring

The previous sections will help to define the scope of service and hours of service that the service provides. It should also unearth information about actual availability and performance. The next step is to add customer feedback to this so that the process of iteration can begin.

It is easy to get over-technical and to try to do too much with Customer Satisfaction Survey. It is therefore suggested that a Customer Satisfaction Survey itself be kept to about four pages avoid too much technical detail. It should aim to score the customer's perception of the quality of each application and service as it is delivered to them. It should also provide the customer perception of:

- availability
- network response / latency
- application response
- ease of use
- help facilities and documentation
- training
- what the hours of service should be

- acceptable downtime
- cost or impact of downtime
- quality of help desk and customer support service.

Especially if this is the first time a Customer Satisfaction Survey has been issued, this is about as much as one could expect a customer to complete by themselves.

Care should be taken with interpretation of the Survey since folklore lingers - especially of poor service. In one case a customer cited "bad" service and, when pressed, his example was of a service that had not been offered for several years.

An example of a Customer Satisfaction Survey follows this Chapter at *Annex One*.

4.9 The Service Management Toolkit

Such tools as those described above are an invaluable aid in service management. A list of web site addresses for performance measurement and monitoring software is at *Annex Three* to this Chapter.

They do not, however, come cheaply. One alternative to the monitoring methods described above is to monitor performance "at-the-desk" by introducing a periodic, programmed test basket of transactions and recording response time for later analysis. This approach is incorporated in some proprietary monitoring tools.

Another possibility is to use a wandering PC with response monitoring software, to sit on the customer's desk during defined monitoring periods.

Whatever monitoring method is used, the service quality has to be measurable, and the measurements have to be demonstrably objective to the customer.

Chapter 2 identified the importance of having Service Management,

Infrastructure Management and Installation Management disciplines in place before embarking on SLAs.

The diagnostic and monitoring tools above add to the Service Management Toolkit (Figure 74.1) with which we can now start to build a SLA.

Figure 4.1: Service Management Toolkit

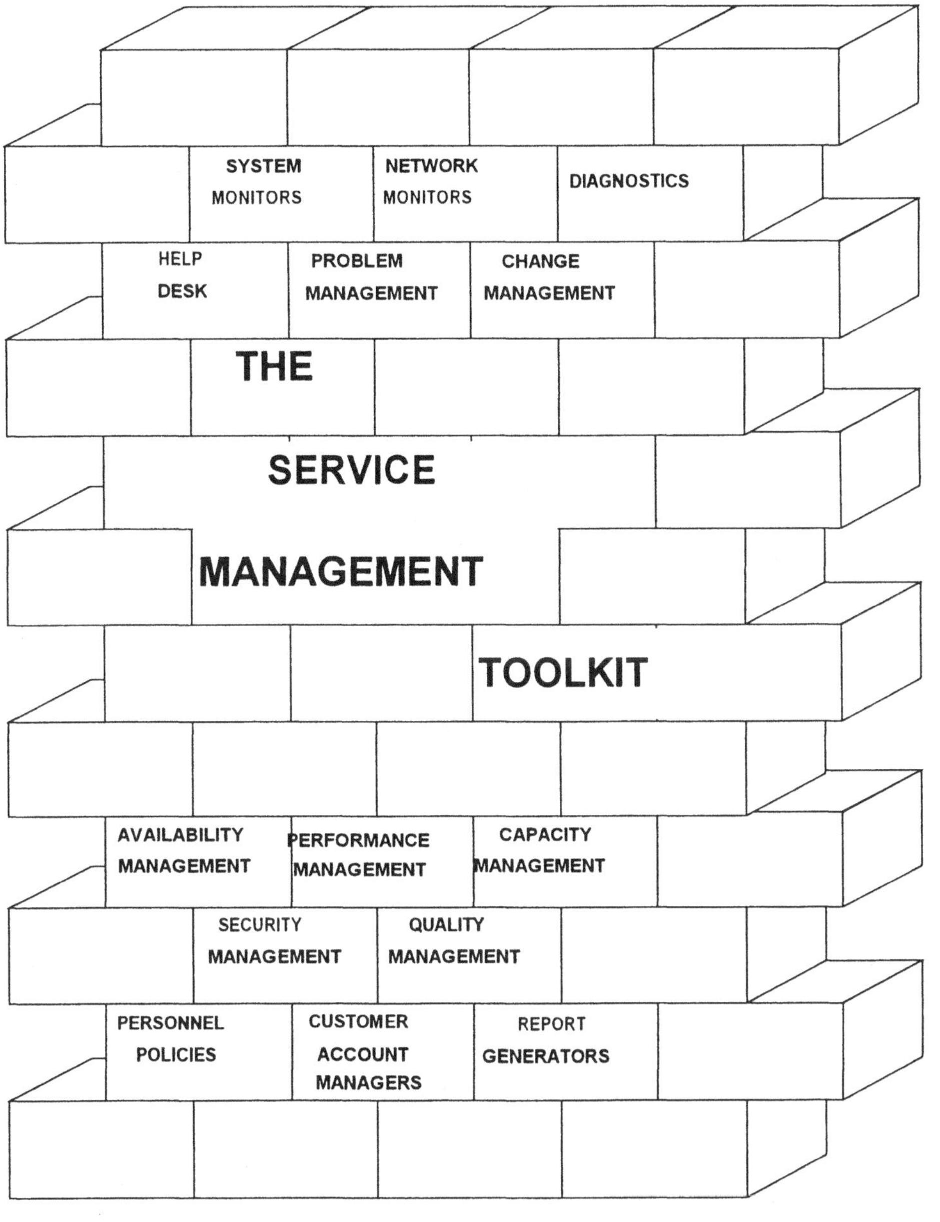

4.10 Monitoring & Litigation

In the event that the service provider fails the contracted service levels, performance monitoring software provides hard, factual evidence to support litigation.

Monitoring the vendor's performance could provide evidence for compensation. Market watcher Iowatch claims that SLA-breaking Internet Service Providers could lose up to $45 million in 2001 if all their customers invoked penalties and damaged for SLA failures.

The key to effective vendor management lies in clearly defining both business and technical requirements in the supplier-customer contract, translating them into measurable service levels and monitoring the achievement of these service levels. This is likely to avoid the need for litigation: but, should it not, it will certainly help your attorneys achieve a just settlement in any subsequent legal action.

4.11 Balancing Detail with Practicality

Given the facilities of sophisticated monitoring software, it is tempting to measure everything that is measurable and to include all of it in a SLA. That is not necessarily what the customer wants.

We have seen how quantification and measurement can help in the context of Critical Component Failure Analysis. This level of detail is of considerable value to the computing service internally to help to introduce alternate paths; to identify where and how to build in processing and network resilience; and to help create the business justification for it. But most of this detailed information is probably not of particular importance to the customer. Too much detail can be indigestible.

Appendix A is a SLA Checklist, developed from one produced by George W. (Bill) Miller for use by American Airlines. It contains some thirteen pages.

The pages consist of only headings and the headings cover most of the items which one could consider (but not necessarily include) when designing a SLA. If one were to include all those headings, the final SLA could run into several volumes!

4. 12 The Balanced Scorecard

One way of simplifying SLAs is to use a balanced scorecard. This summarises service levels and prioritises service levels by weighting them. Failure scores so many adverse points. If the total points score exceeds the agreed threshold, penalties may be invoked. An example is at Figure 84.2.

Figure 4.2: Example of Balanced Scorecard

Support Category	Breach Situation	Points per Breach
Trouble ticket	Failure to issue Trouble Ticket no.	50
Proactive reporting	Failure to advise customer in 1 hr	100
Escalation report major outage	Failure to report major outage in 24 hrs	100
Daily flash report	Failure to provide Mon-Fri by 0900 hrs	100
Capacity Management	Failure to manage capacity effectively	350
Account Management	Failure to attend scheduled meeting	50
Account Management	Failure to provide minutes within 5 days	50
Performance reports	Failure to provide as agreed	200
Change Control	Failure to manage and notify outage	250
MTTR	Failure to arrive within x hrs	100
Implementation date	Failure to meet deadline	400

Penalty & Right to Terminate				
Total Monthly Penalty Points	Call Meeting	Take Action	Invoke Penalty	Cancel Contract
0-1999	No	No	No	No
2000-2999	Yes	Yes	No	No
3000-3999	Yes	Yes	Yes	Site only
>4000	Yes	Yes	Yes	Contract

4.13 What to include in a SLA

Especially when embarking on implementing a SLA for the first time, this checklist approach has much to recommend it. It will ensure key items will not be overlooked and it enforces a consistent approach across different customers. Following the checklist tends to force decisions on what should and should not be included in a SLA and it greatly assists the education process (for both computing service staff and customers).

Broadly, a SLA should cover the following topics:

- purpose of SLA
- duration of agreement
- description of application or service
  - service overview
  - corporate dependence
  - priority
  - critical periods
  - peak periods
  - impact and cost of outage
  - availability
- definition of availability
- availability targets (which may be for the service as whole, or for each component of the service)
- reliability
  - number of incidents of outage
  - packet loss
- transaction rates
  - standard day
  - peak periods
- response / latency requirements
  - or each transaction type
  - definition of response
- forecast utilization and growth/decline
  - now
  - projected growth in 6 months
  - projected growth in 1 year
  - projected growth in 2 years
- batch work details
  - deadlines for input to computing service
  - turnaround targets
  - output arrangements
- storage requirements
  - primary storage
  - secondary

  - archive and retention periods
- accuracy
  - of input
  - of databases
  - of output
- security
  - hardcopy output
  - physical access control
  - logical (systems) access control
  - back-up
  - disaster recovery and contingency planning
- service hours
- scheduled downtime
  - for hardware and plant maintenance
  - for software maintenance
  - for installation or software changes
- support hours
  - help desk
  - technical support
  - customer account management arrangements
- problem escalation procedure
- charging arrangements
- change control
- service level measurement
- monitoring actual service level against SLA targets
- service level reporting
- penalties for failure / bonuses for over-achievement
- arrangements for customer/supplier review meetings
- complaints procedure
- arbitration / mediation / alternate dispute procedure
- termination (including termination on SLA breach)
- handover (possibly to another supplier, on SLA breach or SLA/contract termination)
- contacts
- duration of agreement and renegotiation arrangements
- definitions.

Definitions need to be explicit: what exactly is “availability;” “response;” “respond;” “the network;” “equipment;” “configuration;” “PC;” “move;” “install” mean? What do they include and /or exclude?

For an external supplier, preferably the negative and legal aspects will be dealt with in a contract so that the SLA, as an Appendix to the contract, just details the service specification and service levels. This reduces friction from referring to the contract proper in the event of minor SLA breaches. Appendices to the SLA could deal with detail about charging algorithms, scheduling and discount arrangements for non-prime time and definition of the various regimes under which the service operates. Details of standard services or standard tariffs could also be included as appendices.

4.14 Shell, Template, Model and Standard SLAs

There are many ways of designing a SLA:

Shell SLA

A shell SLA is a standard format with a series of headings like the Checklist at Appendix A. The assumption in using a shell SLA is that each customer’s needs are likely to be unique. Inconsistencies between SLAs can arise from different interpretations of the shell checklist.

Template SLA

A template SLA covers every service type, all service levels and all customer types. It would never be implemented as it stands: much of it would be irrelevant to any specific customer. Its value lies in being the comprehensive service level agreement encyclopaedia from which just the relevant clauses can be drawn for any single customer or service.

Model SLA

A model SLA is intended to be an equitable framework within which the terms will be negotiated for each customer. A model SLA represents the normal conditions of providing service but expects these to be modified in practice.

Standard SLA

A standard SLA is often applied in a commercial bureau environment and is often more of a contractual document. It specifies terms and conditions which would not normally be varied. A standard SLA may be vague on specific performance criteria and heavy on reasons for non-fulfilment of targets since it frequently favours the computing service over the customer.

One-page SLA

A minimalist one-page SLA may be acceptable in some cases. This assumes that A Service Catalog has been developed with Service Categories and Service Products defined as explained in 2.7 above. An example of a one-page SLA is given at Figure 4.3 below.

Figure 4.3: The One Page SLA Format

Service Category: Service Product: Service Owner: Service Profile: Service supplier:	Limits of Service:
Service Definition How to procure: Cost: Service Level Metrics: Availability: Reliability: Response:	Customer Responsibilities:

The steps to creating a one-page SLA are:

- Place services into categories - sections for Catalogue
- List each category as a Service Catalogue section
- Establish integrated / packaged / bundled service products
- Identify modular service products
- Define each service product
- Establish service owner and supplier
- Cover procurement - how, cost
- Specify service levels - availability, reliability, response
- Define limits of service
- Define customer responsibilities.

The Service Definition step involves defining:

- The service itself
- what it provides (the service deliverable)

- where it is provided
- when it is provided
- who supplies it
- how it is supplied
- for whom it is supplied.

Figure 4.4 below provides an example of a completed one-page SLA.

Figure 4.4: The One Page SLA Format

Service Category: Maintenance **Service Product:** Mandatory Engirneering Change **Service Owner:** IT Ops **Service Profile:** Work required by the manufacturer of the equipment to be carried out to the equipment **Service supplier:** ABC Inc	**Limits of Service:**
Service Definition Work completed to manufacturer's specification **How to procure:** Call ABC Inc Help Desk on 00.00.00.00 **Cost:** As quoted by manufacturer **Service Level Metrics:**	**Customer Responsibilities:** 1. Schedule downtime 2. Provide car parking space 2. Provide access to computer room

Availability: 0800-1800 Mon – Fri excluding public holidays **Reliability:** First time fix. No loss of availability following change. Completed within timescale indicated by manufacturer **Response:** Within 48 hours of call	

While each type of SLA has its benefits, for an in-house service, the model SLA will probably be favored.

4. 15 The Service Handbook

A Service Handbook could simplify and supplement SLAs by removing detail, especially transient detail, from the SLA. The Service Handbook could contain:

- Service Mission Statement
- overview of the Service
- hardware configuration
- system software
- standard customer hardware and software supported
- standard service menu
- standard terms and conditions
- default or standard service levels
- standard security arrangements and security advice
- standard contingency planning arrangements (e.g. back-ups)
- standard problem management arrangements (e.g. Help Desk and

escalation procedures)

- technical support details
- scheduled outages for maintenance
- charging
  - standard tariff rates
  - charging algorithms
  - accounting reports
- menu of "gold star" options
- menu of special services
- Computing Services contacts (names and details).

4. 16 Service Level Survey

There was deliberately no technical detail in the Customer Satisfaction Survey, discussed earlier. Unless the customer is very technologically aware, the specification of a service level in technical computing terms will probably be a process of exploration and explanation between a customer and a representative of the computing service, translating business terms (the production of so many cheques, invoices, payslips etc) into computer terms (Megabytes of storage, baud rates, definition of 'transaction' etc).

A Service Level Survey can standardise and formalise this process. If this is completed by a representative from the computing service in discussion with the customer, and jointly agreed, it can be signed off by both and can form the basis of the SLA. More text and detail can then be added.

An outline for a Service Level Survey follows this Chapter at Annex Two.

Many of the components of Service Level Management will now be in place (see Figure 4.5).

Figure 4.5: Components of Service Level Management

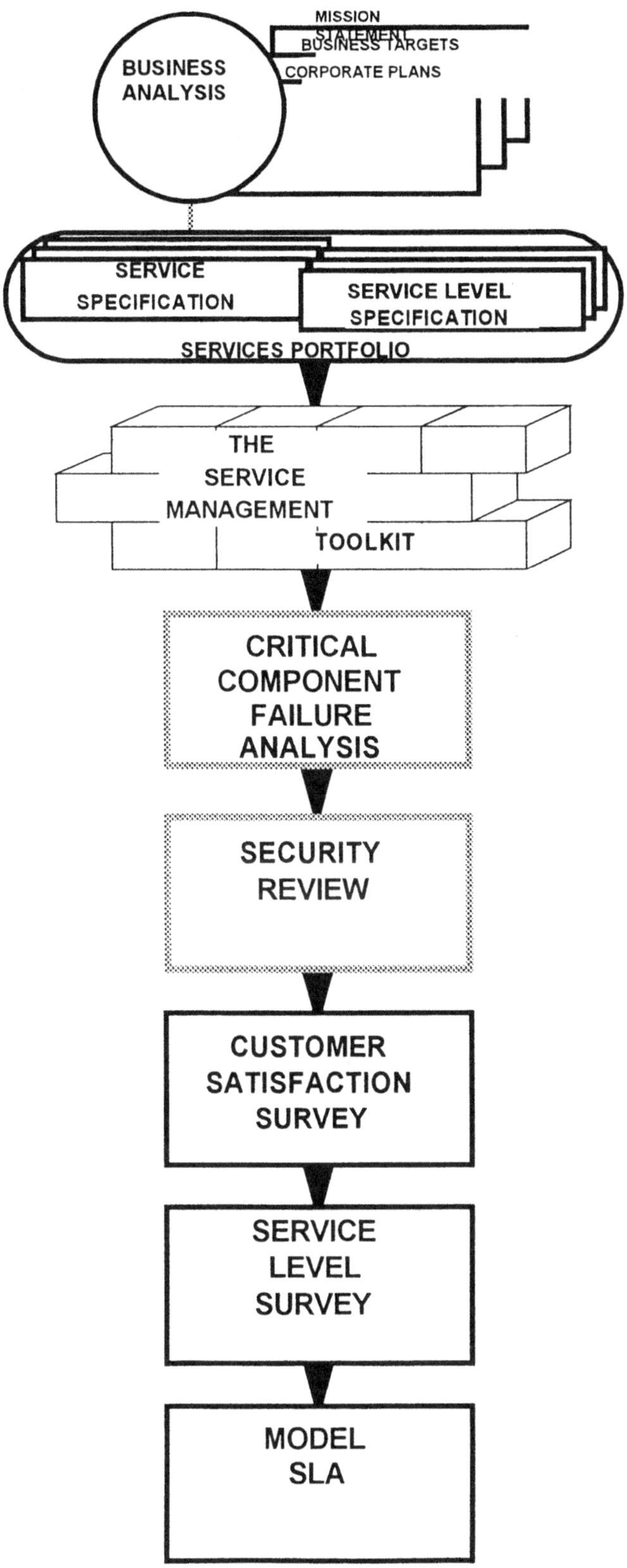

4.17 Charging for Services

The business view of computing is not that it delivers MIPS, MHz or MB, but that it delivers sales reports by area and salesperson, produces bills to send to clients, handles personnel records and so on. We have established that there is a need to translate computing resource usage into a unit understandable by the rest of the business. The most readily understood unit is money!

How else should you measure the computing service except in terms of what it costs, what it saves and what it earns?

Just a few examples of increasing corporate spend on IT will indicate the size of the computing juggernaut:

- over a ten-year period, UK local government IT spend was forecast to increase by over 700%. 31% of councillors said IT departments wasted money. (Source: Gallup & Kudos Research)
- spend on IT represents the largest single area of capital investment in the USA.
- Market leaders are investing heavily in IT
    - retail chain Marks & Spencers in the UK have experienced 200% IT growth in 3 years
    - over 20% of total spend by major banks and financial institutions can be on IT.

But according to Price Waterhouse, IT cost containment is the number four problem worrying management. Chief executives are no longer being swayed by purely technical arguments: they want to see a benefit on the bottom line. Since only users can say whether the IT spend is worthwhile, the solution is seen as charging the user for internal computing services and letting them vote with their budgets.

There are snags with the charging approach:

- charging may just result in under-utilization of corporate assets which have already been paid for
- it costs little more to run a heavily utilised service than it does to run a

lightly loaded service

- unless there is strong central control on desktop and minicomputing, charging may merely duplicate spend on hardware and software as users buy their own equipment which they perceive to be cheaper than using the corporate computing service
- job accounting systems cost in software, system and processing overhead and staff time.

But it may be thought that if SLAs are to be meaningful, the cost of the service should be charged out. So, despite these snags, the decision may be to proceed with charging. If so, a mechanism has to be found to convert resource utilization into cash cost.

There are several possible approaches:

- cost notification
- cost allocation
- charge-out.

Cost Notification

With cost notification, usually the cost of running the computing service is established and users are advised of the amount that is attributable to their use. This is toothless: real charges are not applied and a typical reaction is a shrug and "so what?". However, cost notification is sometimes seen as the first step towards cost allocation or charge-out to give the wasteful customer due warning to become more efficient in the use of computing resources (and to force economies on the computing service) before real bills are imposed.

Cost allocation

Cost allocation identifies the total cost and allocates it, usually according to usage, to customers of the computing service. Since the customers have had little opportunity to establish whether the cost is reasonable, it is often viewed as an inescapable and unjust tax imposed by the unaccountable. Cost allocation frequently causes a sense of grievance with the computing service. Its customers may argue that the computing service can spend what it likes . but it is the customer who picks up the bill for it. But if cost allocation is an interim step to charge-out, the writing is on the wall for the profligate computing service.

Charge-out

A charge-out system bills the customers with the actual costs they incur . and may be accompanied by the right for internal customers to go to alternative sources of supply for computing services. If so, the computing service may be allowed to tout for external work . perhaps on the way to becoming a fully commercial bureau business. The main issues in a charge-out system are defining the charge-out units and the price of them.

At the same time, cost recovery policies have to be set. Most customers would object to a strictly demand-side break-even recovery policy, where the price of a unit could vary according to utilization of the systems. On a lightly loaded system the cost of a job this week could be more than the same job would cost the next week, when the system was more heavily loaded with more users to share the cost. Such a policy penalises the loyal, regular users and complicates their budgetary planning. It should, however, recover exactly the cost of the computing service.

A *standard* costing system would establish the average utilization, create a unit of charge, and base the price of that unit on the average utilization. However, while a breakeven policy should result in balanced books for the computer centre, a standard costing policy could result in either a profit or a loss depending on whether the forecast average utilization was under-achieved, correct, or over-achieved.

The next issue is to establish the unit to be charged out. Most pieces of equipment, and some software, will have resident programs or subroutines to record the usage made of them. CPU time is logged; input/output (I/O) activity and disk storage is recorded.

Similarly tape mounts, pages of output, numbers of microfiche can be identified by job. The utilization of all of these resources can be identified and from this a representative composite Computer Resource Unit (CRU) can be created. The CRU is thus an average of resources used. By examining various types of job, or various transactions, we can identify how many CRUs each one takes. This then gives us another average cost, expressed in CRUs, for each type of job or transaction. This average cost in CRUs can be described as a Computer Workload Unit (CWU). We can now express the cost of providing the computing service in terms of CWUs.

As far as the customers go, the CRU and the CWU are so much technological mumbo-jumbo. The customers are interested in how much it costs to make a ticket reservation or to process an EPOS or ATM transaction. So the next step is to measure the cost of these business activities in terms of CWUs. From this we can create a Business Workload Unit (BWU) which enables us to put a cash cost on each business activity: we can now say what each ticket reservation, each EPOS transaction, or each ATM transaction costs in cash terms.

BWUs have identified hardware (and possibly software) resources per job. General support costs may also be included: maintenance, depreciation, people cost and other items. Support costs specific to a customer may be charged out directly. This charging method is illustrated in Figure 4.6.

A typical example of this technique as applied commercially would be a payroll bureau, which would quote a cost of 'n' cents per payslip.

Figure 4.6. Charging for Computing Services - Schematic

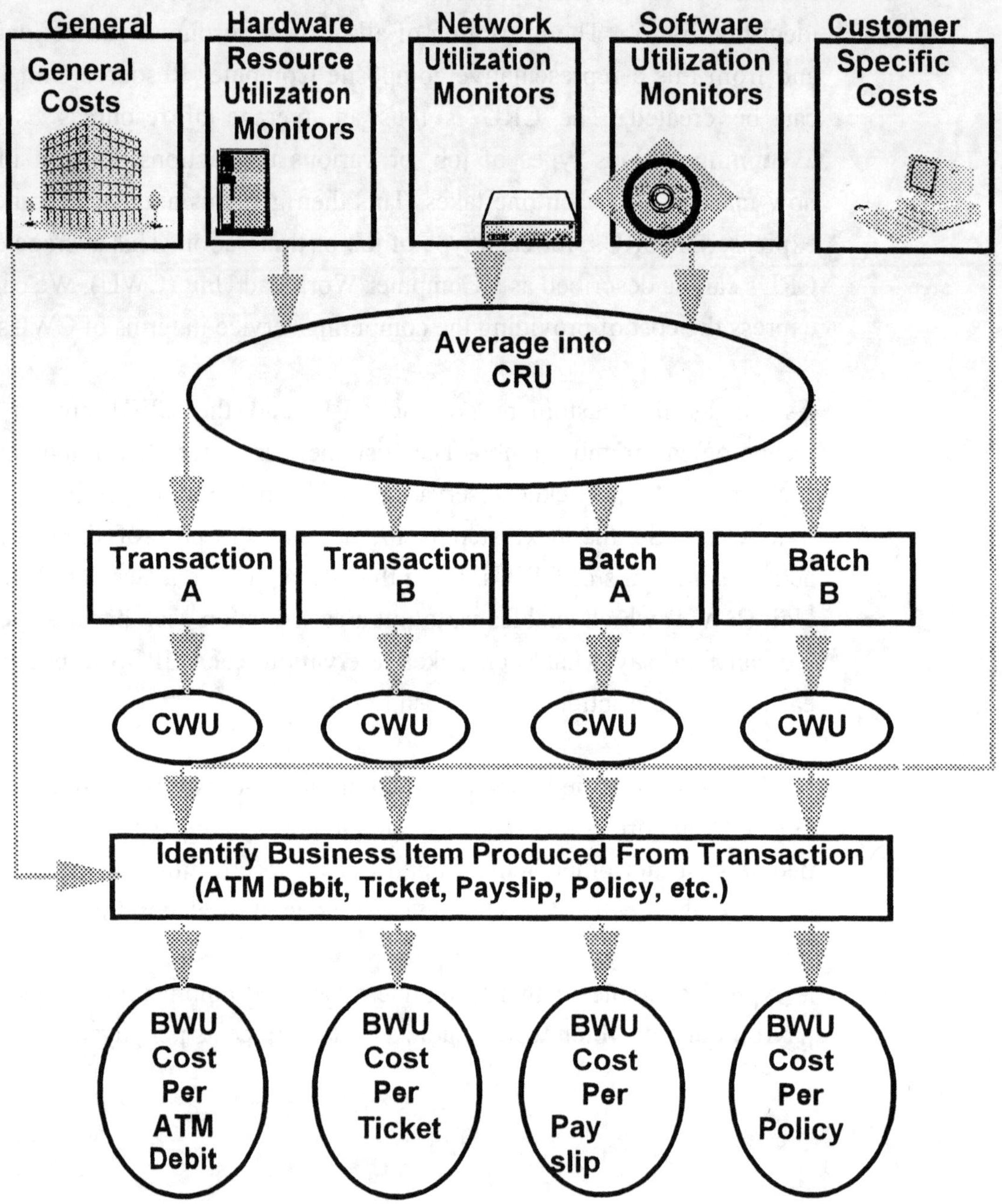

One international consultancy recommends the Service Cost to be identified along the following lines:

Regime A transactions 0.20 cents each.
Regime B transactions 0.25 cents each
Regime C transactions 0.30 cents each

If Service Level Targets specified in this SLA are not met, rebate will be given for each shortfall.

Current use of the system is:

November total journal entries per day				6,400
November total related CICS transactions per day				19,300
Regime A transactions	7,200	@ 0.20p each	=	14.40
Regime B transactions	7,000	@ 0.25p each	=	17.50
Regime C transactions	5,100	@ 0.30p each	=	15.30
	19,300			47.20

Note: Minimum daily cost will be as above
Transactions arriving after 1500 hours will be subject to a surcharge of 0. 10 cents per transaction.

We can vary cost to encourage utilization smoothing . making BWUs dearer in prime time than in non-prime time and perhaps even cheaper at weekends or public holidays. We can create different charging regimes to smooth out demand . and so, perhaps, defer an expensive upgrade. We can offer an "express' service . at a higher cost. Charging therefore offers considerable scope as a capacity management tool . but only if the user is made to commit to using the service for a reasonable length of time.

What is a reasonable length of time? Hardware is usually written off over 3 or 5 years. Computing cost per MIP has steadily dropped .arguably at something like 15-20%/year. New models (or 'kickers' to extend the life of an existing range of hardware) seem to be announced every few months. Maintenance costs on new equipment have similarly dropped as a percentage of capital cost. New equipment is miserly with power compared to older hardware. What that means is that the computing service is saddled with heavy depreciation costs and heavy running costs for the write-off life of the equipment .with high, effectively fixed, costs which can be undercut by a competitor service as soon as a new model or a hardware price reduction is announced. The computing service therefore has one of three choices:

- to maintain the BWU price at broadly the same real cost (allowing for inflation) over the life of the hardware, while competitive computing costs are dropping
- to drop the price for each BWU in line with market forces and to seek operational economies to prevent under-recovery of costs
- to inflate the price for each BWU in the early years and reduce costs in line with market forces later.

From the computing service viewpoint, the ideal is to commit the customer to use the service for the life of the hardware, at a price which will recover capital and operational costs and which will remain broadly in line with competitor services over the life of the hardware.

Unfortunately, the computing service may not be able to determine its own charging policy: frequently this is decided upon by corporate accountants who will not allow over-recovery in the early years to offset future under-recovery as competition strengthens and usage patterns change.

4. 18 Infinite Capacity and 100% Availability?

It was stated earlier that customers are increasingly expecting a utility service, with the computing service available whenever they want it, with whatever capacity they require. The SLA can be the vehicle where the cost-benefit analysis is undertaken on a customer-by-customer, service-by-service basis to establish a balance between cost, availability and capacity that enables corporate targets to be met in the most cost-effective way.

Capacity Cornucopia

Parkinson's Law applies to computer capacity, too: given an excess of capacity, customers' (and the computing service's) usage will expand to fill it.

Complex database enquiries can hit performance and response times and, for some applications, it is not unknown for production of reports to take more processing than the prime application from which they are derived.

Customers are notoriously reluctant to clean out storage and many of the personal computing and office automation facilities particularly electronic mail can flood storage.

The SLA can be used to discourage excessive use of resources. Complex enquiries or reports might, perhaps, be given a lower processing priority or scheduled into a non-prime regime. Exceptional authority could be required to increase the priority or change the regime.

Similarly there might be a standard default time for data to be held on-line: data which had not been accessed could be migrated through managed storage software to slower on-line devices or held off-line. The SLA negotiations could explore these possibilities with customers.

If the computing service charges for services the SLA can include standard costs for standard enquiries and a higher tariff for non-standard enquiries. Equally, the charging policy could encourage the migration of data from disks to automated tape libraries, or to mass storage, or to archive.

Absolute Availability

Does your organization really need 100% availability? Does it appreciate its price? 99.8% availability is relatively easy to achieve these days. 99.9% is trickier, but not impossible. After that it gets harder and costs exponentially more. It cost Bell around 1,000 man-years of effort to improve the availability for computerised telephone switching systems from 99.9% to 99.98% . let alone the hardware cost. For many organizations, the expenditure to achieve a marginal improvement in availability may simply not be cost justified. We get into the laws of diminishing returns. But to another organization, 100% availability may represent the competitive edge that will enable them to capture crucial market share.

The same is true of technical support . especially people. If support is allowed to be entirely demand driven, either the same people get increasingly stretched, with consequent frustration and increase in turnover, or more *staff* are recruited and costs escalate. If charging is applied, support outside standard hours could rate "gold star" (more expensive) tariff.

The SLA process should help to establish the justifiable level of availability for each application or service.

4.19 Realistic Limits to Service

We have seen how SLAs can help to establish the realistic limit service in terms of availability, capacity and performance. Other limits to service may be:

- territorial
  - to define where customers' territory' starts
  - to exclude remote hardware
  - to exclude locations where poor physical or telecommunications access inhibits service provision
  - to exclude locations which computing service resource is unable to support
- service
  - to exclude services beyond the skills or capability of the computing service to support
- technological

- to define 'the configuration', 'the system' or 'the network' on which the services are offered
- to encourage standardization
- to exclude certain types of hardware
- to exclude certain software
- to exclude or limit responsibilities for telecommunications aspects

- logistical
    - to allow for scheduled outages
    - to take account of unscheduled outages
    - to take account of physical difficulties quality
    - to deny or limit service to customers who fail to apply adequate quality assurance or quality control
    - to deny or limit service to applications which have not been developed or tested to quality standards
- security
    - to deny or limit service or access in the interests of security
- functional
- to define each parties' responsibilities
- contingent
    - upon the customer fulfilling certain obligations, typically:
        - compliance with input delivery deadlines
        - compliance with accuracy requirements
        - compliance with volume or utilization forecasts
    - upon suppliers delivering service:
        - internal suppliers
        - external suppliers
    - upon resource (staff, software or hardware) being available.

Many a computing service has been accused of being an ivory tower, but (to mix the metaphors) no computing service is an island. It can only deliver services if services are delivered to it by a number of suppliers. In committing to SLAs, the computing service needs to tie its suppliers to deliver service that will enable the SLAs to be met. Figure 4.7 illustrates the need for back-to-back (also known as tiered or multi-tiered) SLAs with both internal and external suppliers.

Figure 4.7. Back-to-Back SLAs

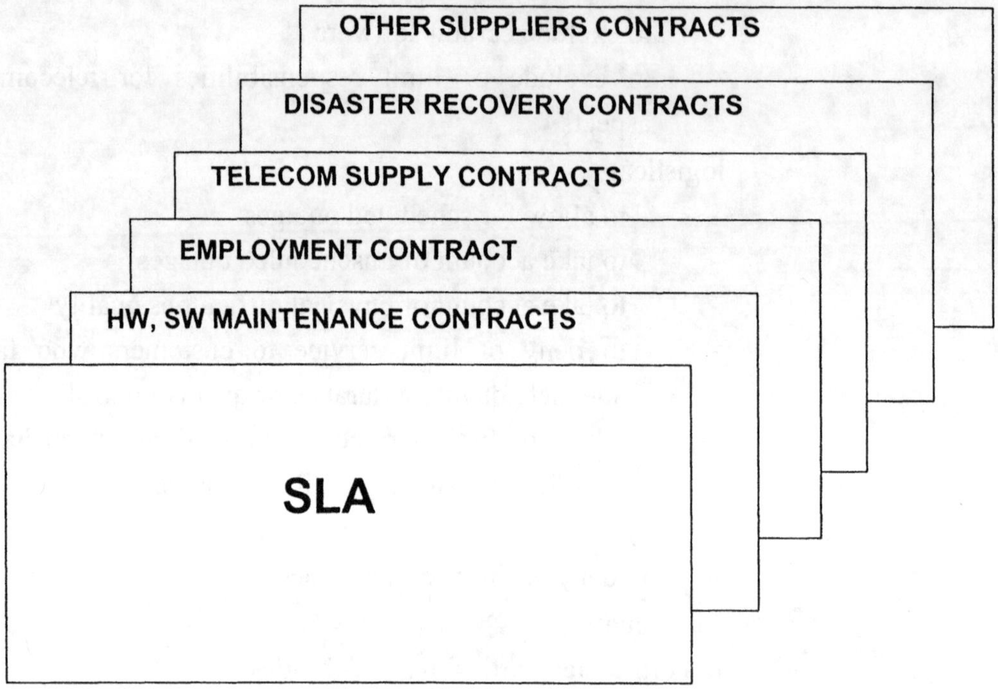

4.20 Penalty Clauses

Loss of availability or slow response can be expensive. (See Figure 4.8 below) and penalties almost always fail to cover the losses incurred to the customer by service outage, since they are typically based on a rebate of the supplier' fees for the service rather than on consequential loss to the customer.

Figure 4.8: Cost of Real-time Service Outages

Application	$000s/hour
Brokerage operations	6,000
Credit Card Authorzsation	2,500
Pay-per-view	145
Home Shopping	110
Catalog sales	110
Airline reservations	100
Teleticket Sales	70
Package shipping	25

Source: Contingency Planning Research & Dataquest

Liquidated damages clauses, while included in many contracts, may not be enforceable if the actual loss cannot be proved.

A good supplier should be prepared to commit to a penalty of up to 100% of its charges a month as compensation for SLA failure. If the supplier is not prepared to accept this, they should be treated with considerable caution. If they do not have confidence in themselves, why should their customers?

As with all SLAs, the service levels should reflect the business need. Maybe the overnight usage is considerably less than, say, 0800 hours to 2000 hours. High availability costs money, and it comes to the point of diminishing returns. Could there be a different service level overnight to extended business day? Do you need 99.999% or better 24 hours a day?

In any event, a supplier will weight the risk of having penalties invoked and will simply budget for the expected penalties in the price charged to the customer. High penalties for relatively minor SLA infringements may simply increase the cost of the service to no great business benefit.

In practice, many suppliers use the contract, Service Level Agreement or Service Agreement to limit the penalties to no more than the fees they charge.

Frequently we find an emphasis on penalties when these penalties are either not enforced or are unenforceable in law.

Liquidated damages clauses are frequently ineffective since, under many legal systems, there has to be proof of loss. It can be remarkably difficult to prove monetary loss, especially where new services are being offered, since there is no track record to project.

The same can apply to situations of rapid growth. Typically a court of law would take a forensic view – that is, it would look at past record and relate losses to that, rather than accept future predictions of income or profitability.

We find that, where penalty invocation is not automatically made by the supplier, most customers do not invoke penalties the first time service level breach occurs – nor even the second time. They are more concerned with getting the service right and with protecting the customer – supplier relationship. This tends not to be the case with public sector enterprises, since their auditors are likely to insist that monies should be recovered.

Equally, although a supplier may be aware that a penalty is not enforceable in law, they may nevertheless make payment in order to retain customer goodwill. Indeed, some suppliers may be aware that they cannot consistently meet service levels to which they commit and may have budgeted for payment of penalties in their bid price.

Consider legal jurisdiction: the country or state in which the law may be applied. Where services cross state, federal or national boundaries, it is important to establish which is the country of first jurisdiction (i.e. where a case will be first heard by a court of law). This may have a bearing both on the interpretation of an Agreement and also on the size of any award made for service failure.

Rather than insisting on service levels that a supplier is likely to fail, in some cases it might be preferable to offer a bonus for over-achievement (providing there is a clear business benefit to this).

4.21 Planning For Change

Both customer and supplier may change their requirements and the SLA should allow for this. The supplier may enlarge the infrastructure; improve SLAs with its suppliers; and, enhance capacity, availability, reliability and latency. The SLA should allow sufficient flexibility for the customer to take advantage of enhanced vendor capability, while the vendor will need an agreed lead time for customer service changes.

Upgrades may cause disruption as well as provide benefits, so it can be a good idea to document, understand and agree the vendor change schedule for new products or upgrades as part of the SLA.

4.22 Organizational Issues

We have seen power shift from the computing service to the user vividly illustrated by two examples:

- most organizations now spend more on end-user desktop computing than they do on mainframe or mini computing
- end-user departments now control the budgets for over *50%* of applications acquisition.

The consequent trend has been to treat the user as a customer who pays for the service used. The computing service organization needs to reflect this: the organization should help to establish or reinforce the new service orientation. Merely implementing SLAs and renaming 'User Support' as 'Customer Support' helps, but is not enough. Full service culture needs to be adopted.

The culture change works best if welcomed and supported from the top and pushed from below with support from the customer. Figure 4.9 shows a model hierarchy for in-company SLA implementation. The same principles apply to inter-company SLAs. Sponsors establish policy and create impetus for SLAs; Executants are involved in negotiation and implementation of them; Monitors ensure compliance with them and report on divergence between service agreed and delivered.

Figure 4.9: Hierarchy for SLA Implementation

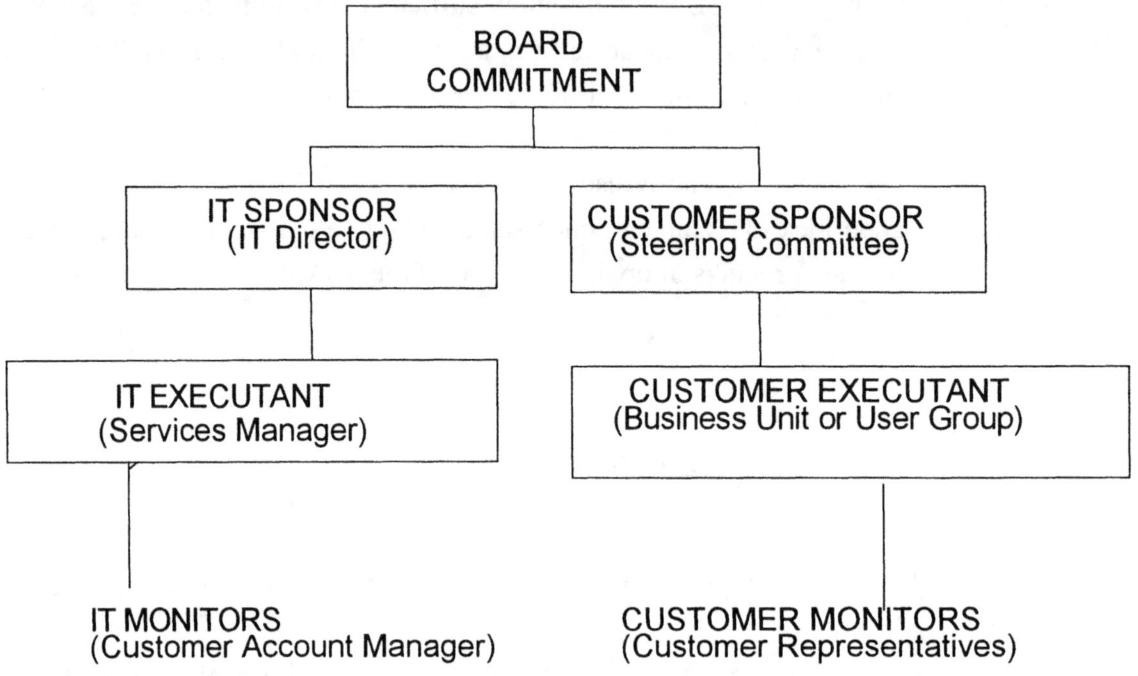

Customer Account Managers

Customer Account Managers may be nominated for each major customer of the computing service. This may not necessarily be a full-time role: in a medium sized installation, or one with relatively few major customers or services, the Customer Support Manager may be able to exercise a Customer Account management function. In a bigger or more complex computing service, the Customer Account Management function could be included as part of the responsibilities of several existing management or Customer Support roles perhaps of the Operations Manager, of Business Analysts or of second level support analysts.

Whether the computing service operates a genuine bureau operation with external commercial customers, or whether it is an in-house service, the role of the Customer Account Manager will be to manage customer accounts to ensure the services provided meet customer needs, and to optimise the customer's use of the computing service. While the Help Desk remains the focal point for customer problems, the Customer Account Manager will be the focal point for the customer to raise general service issues, for monitoring of performance against SLAs and for identifying new opportunities essentially for marketing the computing service's services to that customer.

The Customer Account Manager will also be responsible for co-ordinating all computing service activities necessary to meet the SLA or an agreed customer requirement. Liaison within the computing service is a key part of the job: it involves taking full responsibility for delivery of service to the customer and sorting out any internal problems, procedures or political infighting necessary.

A schematic of the Customer Account Manager role, showing liaison points, is at Figure 4.10. This Figure assumes a classic trident computing service organization. It demonstrates the large number of functions and individuals with whom the Customer Account Manager needs to establish, and maintain, contact.

Figure 4.10: Customer Account Manager: Liaison Points

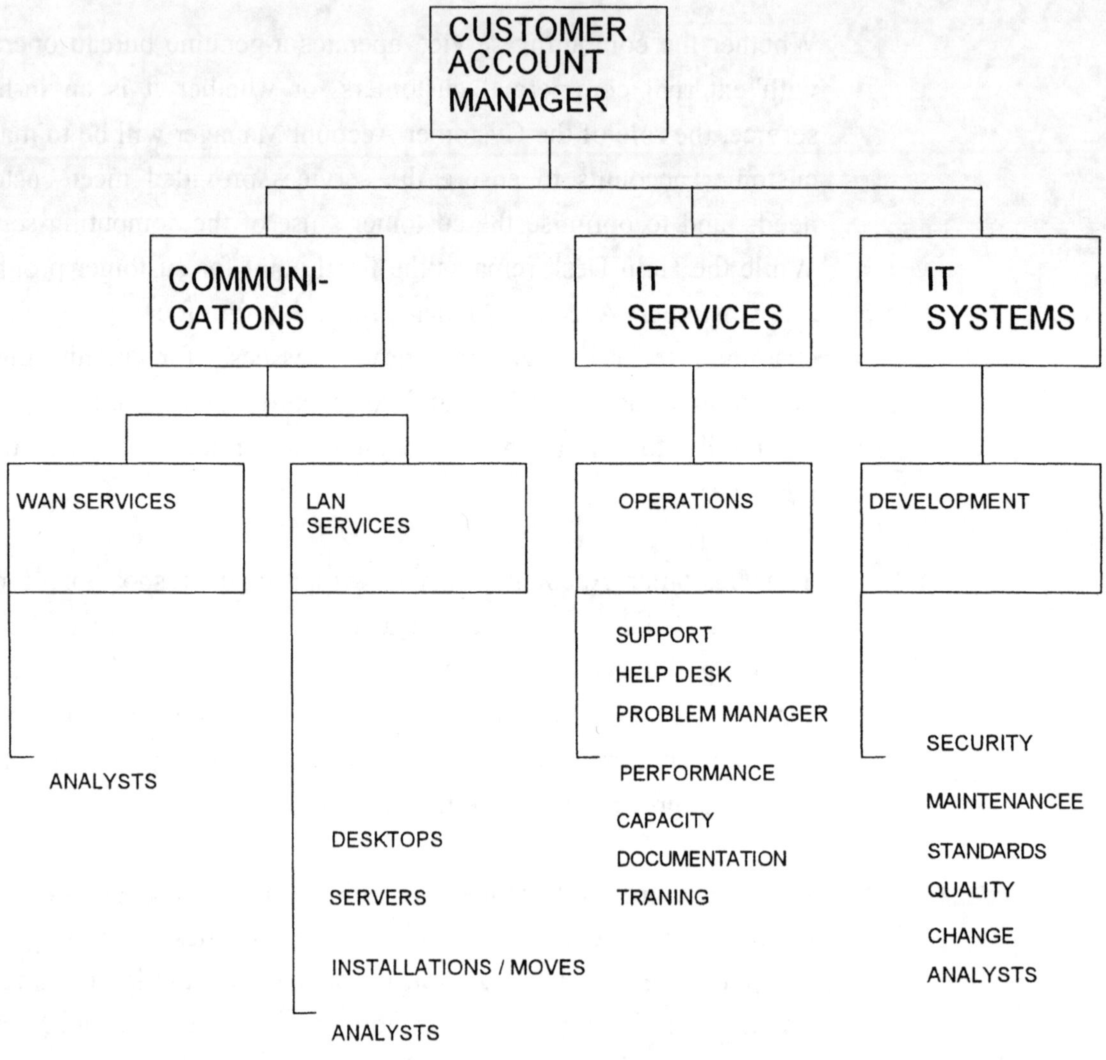

A Marketing-Oriented Approach

Especially where customers are charged and are given the right to find alternative sources of supply of computing services, it may be worth taking this philosophy a stage further - to a full marketing-oriented approach. If the computing service has to be competitive to survive, it needs to equip itself with marketing skills: the appointment of a Marketing and Sales Manager may be a natural corollary.

Example terms of reference for Marketing and Sales Manager and for Customer Account Managers follow this chapter at Annex Four.

4.23 Preparing the Ground

Success in implementing SLAs is less likely if they are just launched on an unsuspecting customer. The ground needs to be prepared first. The ECRIT methodology is designed to address the fundamental marketing, resource and management pre-requisites. ECRIT is an acronym (derived from Education, Commitment, Resource, Infrastructure and Tactics). The ECRIT model for preparation will include:

Education

- of management
- of IT staff
- of customers

by

- raising the topic at customer meetings
- circulating internal papers on SLAs, highlighting benefits
- writing articles on SLAs for in-house newsletters
- preparing presentations for
  - senior management
  - IT management and staff
  - customers
- preparing handouts containing
  - reasons for introducing SLAs
  - identification of benefits
  - sample SLA - sample reports
  - proposals for implementation, monitoring and review.

Commitment

Commitment to the service culture is essential from:

- general management
- IT management and staff
- customers
- suppliers

Resource

- budget
- staff
- equipment and software

Infrastructure

- service management methodologies
- measurement tools
- monitoring and reporting tools

Tactics

- for implementation
  - pilot
  - monitoring, review and modification process
  - to extend SLAs as normal practice

ECRIT says 'Service is written!'

4.24 Pilot Implementation

Before implementing a SLA generally it is well worthwhile trying it out on one specific pilot group. What are the characteristics of an appropriate victim?

The pilot customer should be selected to provide a range and depth of experience of implementing and servicing a SLA. It should contain:

- a fairly large number of users
- a dominant application
- a number of different device types and telecommunications links
- a quality of service requirement
- preferably an existing single point of contact with the computing service
- a reasonable relationship with the Information Services area
- commitment
- a good chance of success!

It is probably helpful if the major activity of the members of the group is either the same application or is composed of a family of applications that form a coherent whole.

It may be possible to test the water with the customer by broaching the subject of SLAs at any existing regular user review meetings. An informal discussion on the principles of SLAs can lead to an agreement in principle to explore the possibilities.

We can use the Checklists and outline SLAs previously discussed to quantify the service requirements of the pilot group and the existing level of service to them.

If we have had to trouble-shoot for this group in the past we may already be familiar with some of their requirements.

A Customer Account Manager can be nominated for them. Success or failure of this pilot may determine whether SLAs are implemented throughout the organization: the choice of Customer Account Manager therefore of vital importance. Tact and interpersonal skills are probably more important than technical skills and the Customer Account Manager for the pilot group might be more senior than would be the case in a ongoing situation.

The SLA hierarchy (Figure 4.9) can be followed to underline and reinforce commitment from both the computing service and the customer.

A pilot exercise should probably last for between three and six months. It may well take six months to implement and see the results of any changes to the service quality. It should be made clear at the outset the there are likely to be changes in either the service or the Service Level Agreement or both during the pilot period: to avoid later embarrassment the SLA should be treated as a draft in that period and not cast in stone.

At the end of the pilot period either the service or the SLA will be adjusted in the light of experience gained. Monitoring tool and method and reporting formats may also undergo change.

4.25 Negotiating with the Customer

The customer's initial response to the concept of SLAs may well be: "At last I'm going to be able to tie down the computing service to actually deliver!" This initial enthusiasm may fade as the customer realises the negotiation is a two-way process and that they are going to have to commit to a certain level of usage, to meet input deadlines and accuracy objectives and all those other targets which are pre-requisite to achieve the computing service levels they wish to establish.

In general, the customer is primarily interested in:

- cost (if the service is charged out)
- availability (only when they need the system, but all the time they need it)
- performance (chiefly response)
- storage (but only to the extent that there should be enough and that access to it should be quick and easy)
- output delivery
- problem solving and escalation
- security (but only when things go wrong!)
- monitoring of performance against SLA.

The negotiation process can only work effectively if these concerns are acknowledged and if it is conducted in terms the customer understands, with a high degree of openness and in a non-adversarial style designed to create a "win-win" situation. For a commercial customer, a bureau is likely to want to protect itself by setting less stringent targets than an in-house computing service might set for in-house customers since their joint aim should be to serve the interests of their business as a whole.

It would be difficult if not impossible to guarantee a service to a single customer without taking account of the needs of all customers: the negotiation needs to get this point across and to identify a fairly short timescale for review of the early SLAs.

The length of time it takes to negotiate a SLA should not be underestimated. If a Template, Standard or Model SLA has already been designed, negotiation with a willing customer could be completed in three or four weeks . maybe even less depending on the complexity of the SLA. For a commercial customer, customers reluctant to embrace SLAs, or for the first pilot SLA, negotiation and implementation may take three to nine months elapsed time. (A SLA similar to Appendix B took over six months elapsed time and some nine man-months effort). There can be a long lead time and high resource cost in implementing SLAs.

4.26 Reporting Actual Performance Against SLA

Periodic reports on achievement or non-achievement of SLA targets are necessary. The reporting timescale needs to be short enough not to average out significant under-performance. For most IT services, service levels are reported on a monthly calendar. This means that different quality may be delivered in a 28-day as opposed to a 31-day month (where a service is constantly monitored, the shorter the reporting period the more accurately will the report reflect actual user experience).

In many services, averages of response or latency are taken by periodic snapshots. Frequently this interval is 15 minutes. However, if response was extremely slow in between the samples, this would not be identified by the sampling. The same is true of availability when sampling is used in this way: unavailability of less than 15 minutes would not be identified.

Figures 4.11a to 4.11c show how an apparently acceptable monthly service achievement is actually unacceptable if measured on a daily basis.

Statistics Distort Actual Performance Over Time.

Figure 4.11a: Monthly Report

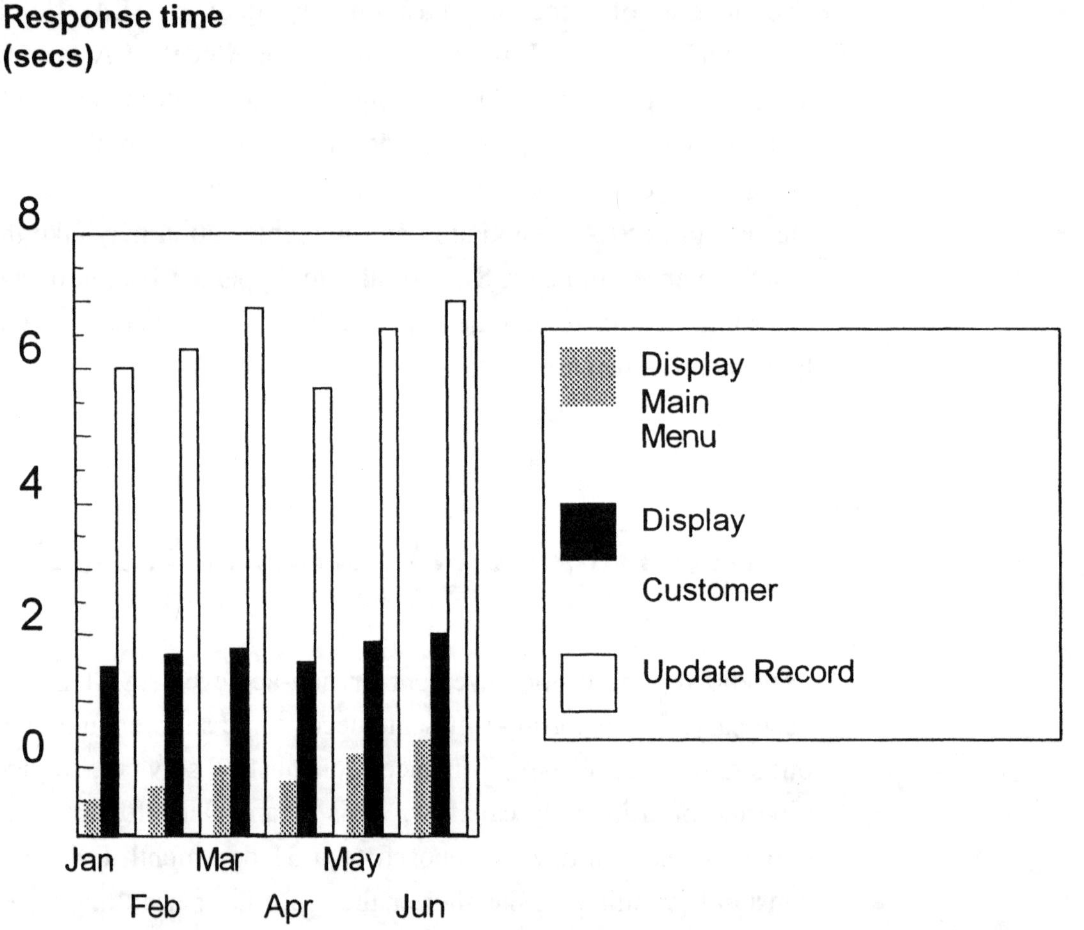

Figure 4.11 b: The Same Data, Weekly Report

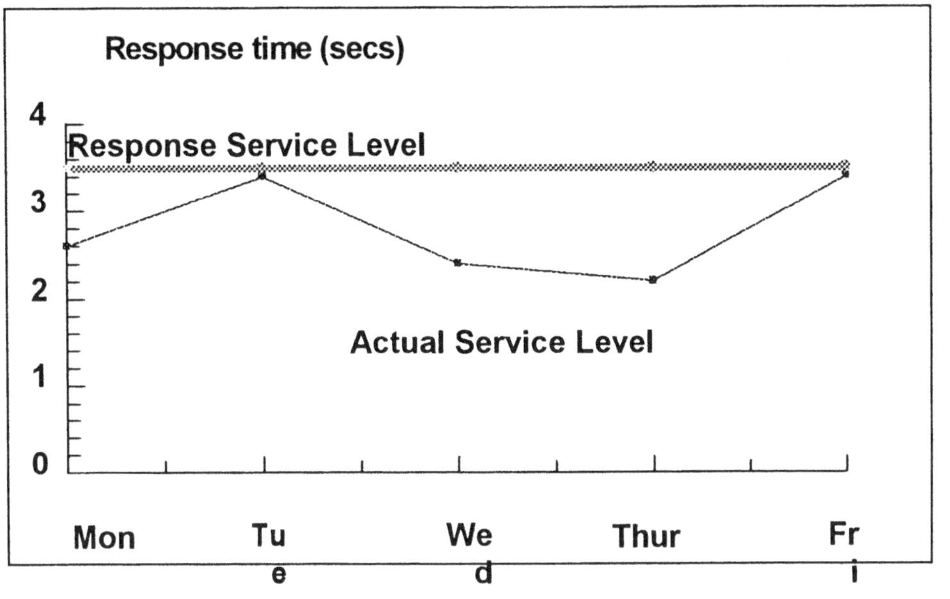

Figure 4.11c: The Same Data, Daily Report

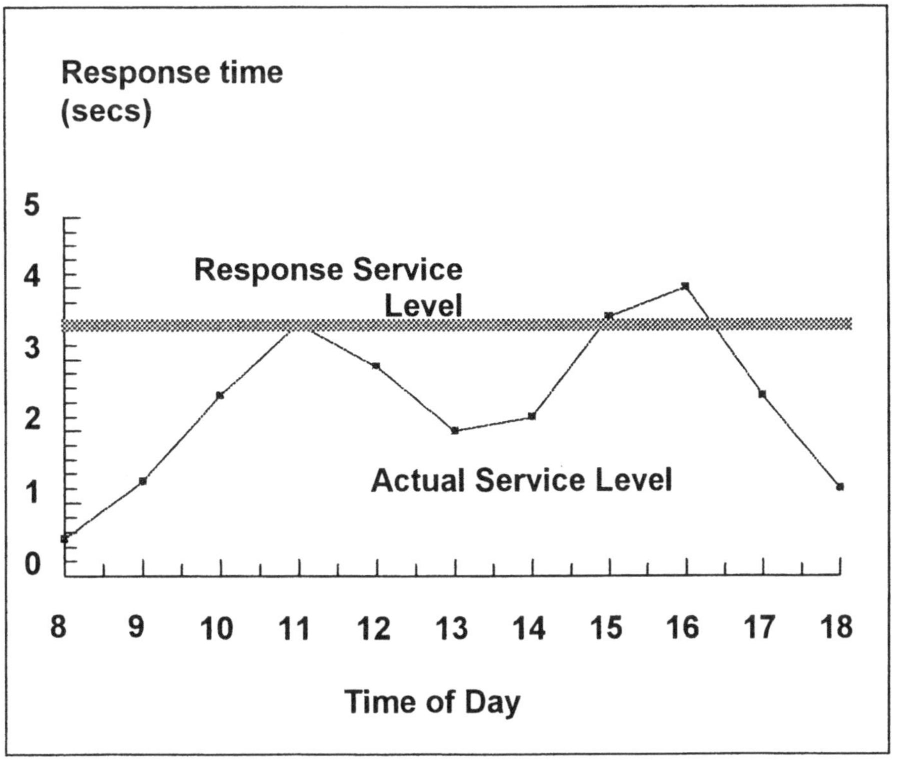

However, service reporting periods do need to be long enough not to deluge computing service and customer in a stream of information. The more common options are:

- each accounting period
- each calendar month
- rolling 4 weeks
- rolling 8 weeks.

A rolling reporting period, while more difficult to handle, provides a better reflection of user's experience than one with a cut-off date.

Reporting is aided by use of the Service Database of performance information. If all incidents and problems resulting in service outages have been reported into the Service Database, it is relatively simple to produce SLA reports from it. However, the Service Database may have to be supplemented with performance information about data preparation, batch output, technical support and other services. The schematic at Figure 4.12 shows a structure for SLA reporting founded on a Service Database:

- on-line service outages are reported direct to a mainframe Service Database from Incident and Problem reporting systems and from system monitoring software
- batch deadlines missed are input to a mainframe or PC-based package
- support services targets missed are input to the mainframe or PC based package
- reports are provided to Operations and to customers.

Figure 4.12. SLA Reporting Schematic

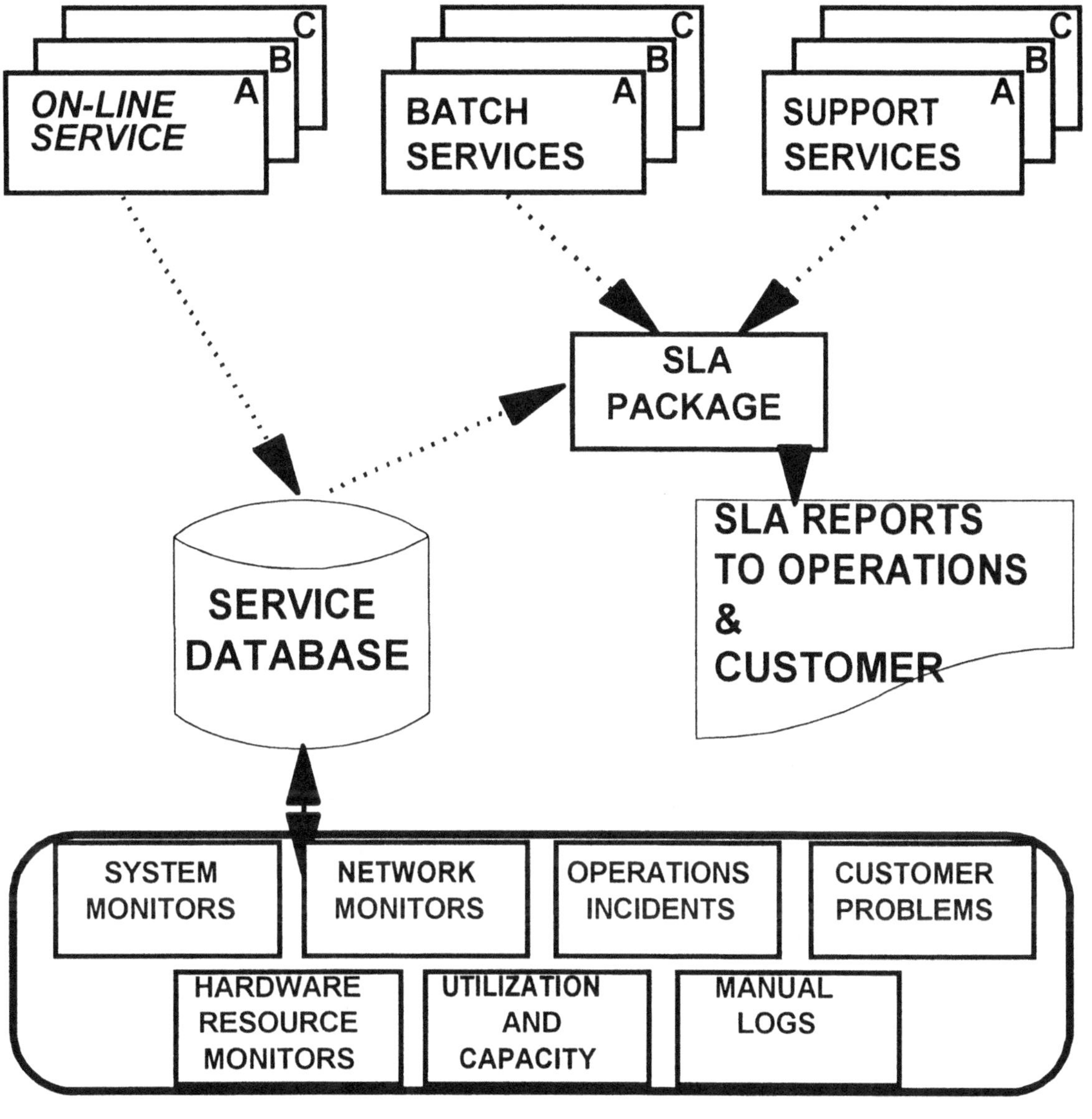

The shape and content of reports will vary depending on the type of SLA chosen. A look at a few examples of reports that are actually in use may help to show the variety of report types which can be used.

A SLA report could cover actual performance against targets in the following areas:

- availability

- on-line response times
- batch turnaround
- output arrangements
- throughput (volumes)
- Help Desk and customer support activity
- technical support.

In the event of any shortfall, the report should identify:

- time and duration of each failure
- number of failures
- reasons for failures
- similar information for partial failures.

At its simplest, if no batch work is involved, global uptime and service availability split between prime and non-prime time may be adequate. (See Figure 4.13).

However, there is a temptation to burden the customer with such reports, which are probably more meaning to IT management than to the customer.

As noted above, the customer views services horizontally. The Human Resource Director's view may be: "The HR service includes elements of mainframe, WAN, LAN, client-server, applications and desktop availability. I do not care where it comes from: irrespective of platform, architecture, or geography, what is the end-to-end availability of the Human Resources service? " IT, typically, views the service vertically: in terms of platforms.

Figure 4.13: Global Service Report – Schematic

19 SAS(R) LOG   Os SAS 5.18   JOB BRDIBM90 STEP PRTPLT   PROC SAS   11:54   NOVEMBER 23. 1890

IBM   SYSTEM AND USER AVAILABILITY UPTIME FOR XXXX   MONTHLY BREAKDOWIN

MONTH	ELAPSED TIME	SYSTEM UPTIME HOURS	SYSTEM UPTIME PERCENT	USERS UPTIME HOURS	USERS UPTIME PERCENT TOTAL	USERS UPTIME PERCENT 8 – 6	SYSTEM BREAKS	SYSTEM DOWNTIME HOURS	IBM ENG HOURS	SERVICES TIME HOURS	S/A S/W TIME HOURS	SPECIAL UNATTRB TIME	OPERATIONS TRAINING TIME
JANUARY	744.00	740.40	99.52%	727.71	97.81%	99.11%	3	3.60	0.00	0.00	1.08	0.00	11.61
FEBRUARY	696.00	696.00	100.00%	680.37	97.75%	99.92%	0	0.00	2.00	0.00	0.00	0.00	13.63
MARCH	744.00	742.42	99.76%	726.74	97.47%	99.30%	2	1.58	7.00	0.00	0.00	0.00	8.68
APRIL	720.00	717.25	99.62%	700.43	96.90%	99.63%	4	2.75	5.58	0.00	5.66	0.00	5.58
MAY	744.00	738992	99.45%	725.18	96.92%	100.00%	1	4.08	2.08	0.00	0.00	0.00	11.68
JUNE	720.00	719.78	99.97%	690,25	96.84%	100.00%	1	0.22	5.80	11.75	0.00	0.00	12.28
JULY	744.00	736.00	98.82%	724.04	96.24%	98.77%	3	8.00	2.58	0.00	0.00	0.00	9.38
AUGUST	744.00	726.31	97.62%	700.53	91.78%	100.00%	4	17.69	4.42	10.32	3.95	0.00	7.08
SEPTEMBER	720.00	716.17	99.75%	705.22	97.69%	99.13%	1	1.83	8.50	0.00	2.07	0.00	5.38
OCTOBER	528.00	527.16	99.84%	521.56	98.62%	99.21%	2	0.84	0.48	0.00	1.83	0.00	3.29

IBM   SYSTEM ANO USER AVAILABILITY UPTIME FOR XXXX SUMMARY BREAKDOWN WITH PERCENTAGES.

HOURS PERCENT HOURS PERCENT PERCENT TIME TIME HOURS TIME BETWEEN TIME TIME RECORD 8 - 6 HOURS tOURS HOURS FAILURES

DATE OF LAST	ELAPSED TIME	SYSTEM UPTIME	SYSTEM UPTIME	USERS UPTIME	USERS UPTIME	USERS UPTIME	SYSTEM BREAKS	SYSTEM DOWN	IBM ENG	SERVICES TIME	S/A S/W	MEAN TIME	SPECIAL UNATTR	OPERATIONS TRAINING
XXXX	7368.00	7327.4	99.45%	7161.39	96.64%	99.52	21	40.55	36.14	22.06	14.59	349.92	0.00	53.21
PERCENTAGES	100.00%	99.45%		97.20%				0.55%	0.49%	0.30%	0.20%		0.00%	1.27%

MEAN TIME BETWEEN FAILURE FOR THE PREVIOUS 26 DAYS 672.00 HRS

The sample SLA reports at Figures 4.14 take another approach. These refine reporting to specific systems and services and report on achievement and non-achievement of SLA targets, but they do not differentiate between critical, prime and non-prime-time availability. They do, however, cover partial outages and cross-refer to a problem management system.

Figure 4.14a: Sample SLA Report

SERVICE LEVEL REPORT FOR APPLICATION: "SYSTEM A" WEEK COMMENCING: DD/MM/YY

THE FOLLOWING IS THE SERVICE LEVEL PERFORMANCE EXTRACTED FOR THE ABOVE APPLICATION. PLEASE ADDRESS ANY QUERIES TO THE PRODUCTION MANAGER, TELEPHONE 12345.

APPLIC-ATION	SLA TARGET	ACTUAL AVAIL-ABILITY	SLA MET Y/N	ERROR:DATE/TIME:OUTAGE PROBLEM
System A	96.00	95.91	No	OUTAGES CURRENT Software ddmmyy 10.11 :36 73422 Software ddmmyy 10.48 :20 73412 Software ddmmyy 10.31 :36 73422 THE FOLLOWING ARE PARTIAL OUTAGES: Hardware ddmmyy 11.20 0:40 73483

Figure 4.14b: Batch Service Level Report

APPLIC-ATIONS & REPORTS	MON	TUE	WED	THU	FRI	TOTAL	ALLOWED	ALL	Y/N
	DEADLINES MISSED								
REPORTS A	-	-	-	-	1	1	2	15	Y
VAN DESPATCH	-	-	-	-	-	-	1	60	Y
REPORTS B	-	-	-	-	-	-	3	40	Y
BATCH 1	-	-	-	-	-	-	1	5	Y
REPORTS C	-	-	-	-	-	-	1	5	Y
TAPE AAA	-	-	-	-	-	-	1	5	Y
INPUT X	-	-	1	-	-	-	1	5	Y
BATCH 2	-	-	1	-	-	1	1	45	N
REPORTS D	-	-	-	-	-	-	1	5	Y
INPUT Y	-	-	-	-	-	-	1	5	Y
OUTPUT AA	-	-	-	-	-	1	1	5	Y
OUTPUT BB	-	-	-	-	-	-	1	5	Y
REPORTS E	-	-	-	-	-	-	1	5	Y

The graph at Figure 4.15 provides a method of measuring a particular service component but the method could equally well be applied to measurement of an entire service, with separate graphs for critical, prime, non-prime times.

Another way to measure IT's overall attainment of SLAs is to create a form of Availability Index (similar to a Financial Times AU Share Index). The simplest way would be to calculate:

$$\frac{\text{(multiple of all actual availabilities for period)}}{\text{multiple of all target availabilities for period)}} \times 1000$$

e.g., for 4 SLAs

$$\frac{99.3 \times 98 \times 99.1 \times 94.5}{99 \quad 97 \quad 98 \quad 95} = 1019.4$$

This figure could be used in trend graphs, newsletters or as staff achievement targets for Management By Objectives (but should over-achievement on some SLAs compensate for under-achievement on others?).

Figure 4.15. Computing Center - Mainframe Availability, 0800 to 2000 hours

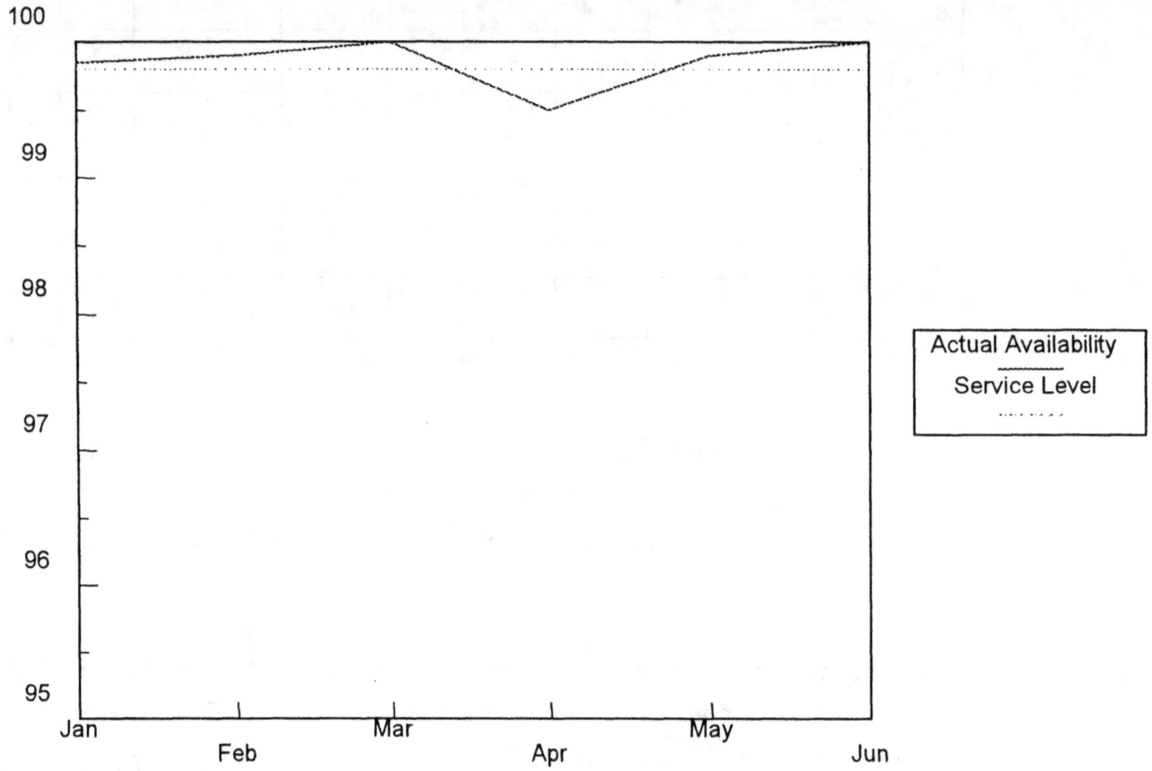

Other ways of reporting include traffic lights, showing red, amber or green conditions for the service. If they have to be photocopied in black and white, the meaning can get lost. Smiley icons can also be used.

It is now commonplace to post service level reports on the Internet, Intranet or Extranet as appropriate.

4.27 Service Review Meetings

Service Level data should be available for discussions at Service Review Meetings. Typically, separate Service Review Meetings would be held with external:

- customers (jointly or individually)
- suppliers (chiefly maintenance vendors, bureau services or contractors) and with internal:
    - customers
    - suppliers
    - accountants (concerning budget, cost and recovery)
    - problem management teams
    - change management teams
    - capacity planning teams
    - customer support and technical support teams.

These meetings will all seek to meet service level targets and help to establish what steps need to be taken in the event of shortfall (or substantial over-provision) of service quality or in the light of changing business needs . whether to adjust the service or to adjust the SLA.

4.28 The Customer Review Meeting

The frequency of meetings and seniority of attendees at the Customer

Review Meeting will depend upon:

- how mature the SLA is
- whether SLA targets are consistently being met
- the commercial value of the customer
- whether the SLA is due for renewal
- whether the customer wishes to negotiate changes to the SLA.

The normal attendees would be the Customer Account Manager (supported by representatives from Operations, Telecommunications, Systems and Technical Support areas and by the Marketing and Sales manager as appropriate) and the Customer Representative (supported by appropriate end-users and specialists).

A typical agenda might cover:

- minutes of previous meeting
- new matters arising
- review of Service Performance against SLA (Hardware, Software, Telecommunications and Support)
    - availability
    - performance
    - utilization
    - batch work
- report on Help Desk & Customer Support activity
- review of outstanding problems
- administrative Issues
- planning
    - proposed hardware, plant, software, telecommunications changes
    - requests for new services
    - service changes (e.g. Public Holiday arrangements)
- adjustments to service or SLA
- any other business
- place, time and date of next meeting.

4. 29 Service Motivation

Service orientation does not just happen: individuals need to be motivated to make it happen. The organizational culture may dictate the tools available to achieve a full service culture. Possible motivators could include:

- setting service targets as Management By Objectives (MBO) objectives and reporting on achievement under annual appraisal systems
- linking performance pay to achievement of SLA objectives
- linking a group bonus to achievement of SLA objectives to encourage

peer pressure against the less service-oriented individuals

- using service-orientation posters of the "Customer is King" type
- immediate incentives for achievement (e.g. bottle of champagne for perfect service record for a month; league tables)
- making conduct which could jeopardise provision of service "gross misconduct" that is, a *firing* offense
- putting financial penalties on the computing service for non-achievement of SLAs (e.g. free processing, so that computing service budget targets may be missed in the event of serious non-performance with consequent post-mortems)
- putting financial or service quality penalties on customers who fail to deliver their forecast usage or fail to meet their side of the SLA
- instituting Quality Circles or other techniques to improve service quality.

4.30 Extending SLAs

When the period set for trialing the pilot SLA is complete and the model SLA has been fine tuned, SLAs can become normal practice for new customers and can be retro-fitted for existing customers perhaps at the time charges are renegotiated or when budgets are renewed.

The components of Service Level management will now be in place (see 4.13).

Figure 4.13. Components of Service Level Management

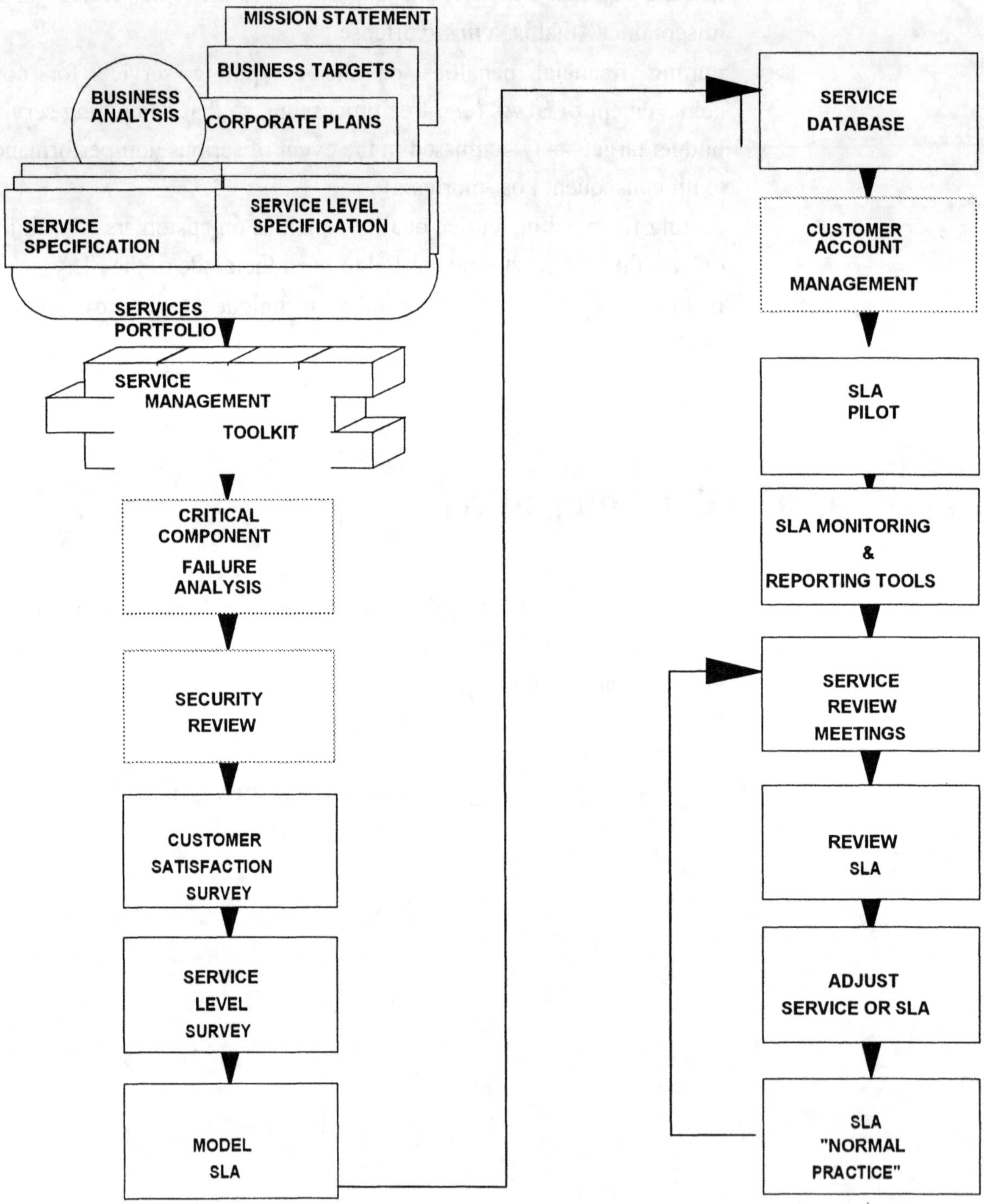

Annex One: Example Customer Satisfaction Survey

CUSTOMER SATISFACTION SURVEY

In order that we may assess and improve the service offered to our customers and users, we would be grateful for your help and opinions as we complete this following questionnaire with you.

Customer Details

Name:

Job function/title:

Department:

Address:

Tel no:

Fax:

E-mail:

Your main contact with the supplier is:

Name:

Tel. no:

PRODUCT:
Please give an average rating for the internet services or products you use on a scale of 1 (poor) to 6 (excellent). Please Use a separate sheet for each product. For each characteristic, please allocate a priority weighting – these weightings should add up to 100.

Characteristic	Priority	Rating					
Service availability		1	2	3	4	5	6
Response time		1	2	3	4	5	6
Reliability		1	2	3	4	5	6
Site design		1	2	3	4	5	6
Ease of use		1	2	3	4	5	6
Cost		1	2	3	4	5	6
Value for money		1	2	3	4	5	6
Documentation		1	2	3	4	5	6
Customer Training		1	2	3	4	5	6
Administration		1	2	3	4	5	6
Other (specify)		1	2	3	4	5	6
Other (specify)		1	2	3	4	5	6

What are the trends in Service Quality over the last (Please circle one)

12 months?	**Improving**	**Stable**	**Declining**
3 years?	**Improving**	**Stable**	**Declining**

Are you satisfied with the service? **Yes** **No**

What improvements would you suggest?

What priority would you place on these improvements? For each improvement, please place a priority from 1 (low) to 6 (high).

CUSTOMER SUPPORT:
Please rate the following characteristics of the Customer Support service on a scale of 1 (poor) to 6 (excellent). For each characteristic, please allocate a priority weighting – these weightings should add up to 100.

Characteristic	Priority	Rating					
Help Desk effectiveness		1	2	3	4	5	6
Availability of support staff		1	2	3	4	5	6
Access to specialist staff		1	2	3	4	5	6
Help with problem solving		1	2	3	4	5	6
Promptness of response to queries		1	2	3	4	5	6
Effectiveness of response to queries		1	2	3	4	5	6
Other (specify)		1	2	3	4	5	6
Other (specify)		1	2	3	4	5	6
Other (specify)		1	2	3	4	5	6

What are the trends in Customer Support Quality over the last (Please circle one)

12 months? **Improving Stable Declining**

3 years? **Improving Stable Declining**

Are you satisfied with Customer Support? **Yes No**

What improvements would you suggest?
What priority would you place on these improvements? For each improvement, please place a priority from 1 (low) to 6 (high).

DEVELOPMENT:

Please rate the following characteristics of the Development service on a scale of 1 (poor) to 6 (excellent). For each characteristic, please allocate a priority weighting – these weightings should add up to 100.

Characteristic	Priority	Rating					
Availability of support staff		1	2	3	4	5	6
Understanding of business issues		1	2	3	4	5	6
Access to specialist staff		1	2	3	4	5	6
Timescale for application delivery		1	2	3	4	5	6
Functionality of development application		1	2	3	4	5	6
Software defects		1	2	3	4	5	6
Other (specify)		1	2	3	4	5	6
Other (specify)		1	2	3	4	5	6
Other (specify)		1	2	3	4	5	6

What are the trends in Development quality over the last (Please tick one)

12 months?	Improving	Stable	Declining
3 years?	Improving	Stable	Declining

Are you satisfied with the Development service? Yes No

What improvements would you suggest?

What priority would you place on these improvements? For each improvement, please place a priority from 1 (low) to 6 (high).

OUTAGE TOLERANCE:

In the event of a major outage, please specify your priority applications and/or service.

Application / Service	**How often used**			**Max. acceptable downtime**		**Average actual downtime**	
	Total[6]	Prime time [12]	Non-prime	Normal working	During Critical periods *	Normal working	During Critical periods

* What do you understand by Critical periods in the context of your work area?

What is the Service Quality during these critical periods?

Rating 1 (poor) to 6 (excellent)	1	2	3	4	5	6

What could we do to improve the service during these critical periods?

For each improvement, please place a priority from 1 (low) to 6 (high).

Please specify any new service or facilities that you feel would improve the service already provided.

For each new service or facility, please allocate a priority from 1 (low) to 6 (high).

[6] e.g. continuously, daily Monday-Friday, 7 days a week, once a week, etc.
[12] if applicable specify periods of 'peak' usage

Annex Two: Example Service Level Survey

Service Level Survey

In order that we may assess and improve the service offered by the Computing Centre to our customers, we would be grateful if you would assist your Customer Support Manager to complete the following questionnaire. A copy will then be forwarded to you for your records.

Customer Details

Initials

Surname:

UID:

Site / Company:

Address:

Telephone No: Fax:

Date:

Computing Center Customer Support Manager Details

Name:

Address:

Telephone No: Fax:

Name of Application/Service:

Transaction Rates and Response Times

Transaction Name?

Transaction Rates	Rate	Time/Day
How many per standard day?		
Per peak hour of day?		
Per peak hour of week?		
Per peak hour of month?		
Per peak hour of year?		

Required response for this transaction type? (alter percentages as required)

90%	<	secs/mins
95%	<	secs/mins
100%	<	secs/mins

	+ 6months	+ 1 year	+ 2 years
Projected % Growth			
Projected % Decline			

Other comments

(Please complete separate form for each type of transaction)

Batch Jobs

	Day	Time
Input to Operations by:		
For turnaround by:		
	Specify	
Output to:		
Despatch Method:		
Security requirements:		
% Missed deadlines acceptable:		

	+ 6months	+ 1 year	+ 2 years
Projected % Growth			
Projected % Decline			

Other comments

(Please complete a separate form for each batch job)

Storage requirements (in megabytes - MB)

(Note: this will be completed by Computer Services based on business requirements specified by the customer)

Primary storage	
Secondary storage	
Other storage	
Archive	

	+ 6months	+ 1 year	+ 2 years
Projected % Growth			
Projected % Decline			

Other comments

Thank you for completing this survey.

Signed:

_________ ______________________________
Customer

____
Customer Support Manager

Date:_______

Annex Three: Terms of Reference for Marketing & Sales Manager and Accounts Manager

The Marketing and Sales Manager is responsible for the overall effectiveness of the Marketing Section. The Marketing Manager monitors that Customer Account Managers are undertaking the appropriate tasks assigned to them. The Manager will undertake the following:

1. Review the Market to identify prospective clients
2. Seek to develop work from existing clients in conjunction with Customer Account Managers
3. Public Relations
4. Based on targets set for Customer Account Managers, set targets for each marketing activity
5. Compute results of marketing activities against targets
6. Ensure Service Level Agreements (SLAs) are negotiated for each customer; monitor delivery of service against SLAs and provide details to Customer Account Managers
7. Monitor the Customer Account Managers' progress.

TERMS OF REFERENCE FOR CUSTOMER ACCOUNT MANAGERS

1. Customer Account Managers will be responsible for monitoring the success of the Computing Service in providing their customers with the services they require in accordance with Contracts and Service Level Agreements. They will also be responsible for identifying customer needs and relaying those to executive areas of the Computing Service; identifying areas of customer's concern and initiating action to resolve the issues which give rise to that concern; identifying customer needs and relating these to services which the Computing Service can provide, thus expanding business with that customer.

2. Customer Account Managers will undertake the following activities:

(1) By personal initiatives and by working with the Marketing and Sales Manager, to promote Computing Services so as to assist the Marketing and Sales Manager close sales.

(2) Identifying and advising the Marketing and Sales Manager of any possible new areas of business.

(3) Identifying areas of industry research required and passing these to the Marketing and Sales Manager for action.

(4) Monitoring service to the customers against contracts or Service Level Agreements (SLAs); identifying and resolving any shortfall.

(5) Management of customer demand as far as possible so as to avoid peaking of work for the Computing Services.

(6) Reviewing customer usage against contract (or SLA): identifying over or under fulfilment and informing the Computing Service Manager of its implications; liasing with the customer on under or over fulfilment against contract or SLA.

(7) Regular visits to customers to a schedule agreed with the Marketing and Sales Manager.

(8) Interim telephone calls to customers between visits to identify and defuse any areas of concern and to promote the further use of the Computing Service.

(9) Arranging for visits to the Computing Service by prospects and customer users for discussions on applications etc.

(10) Arranging visits to the Computing Service by customers to use facilities available.

(11) Co-ordination of the Computing Service resources to facilitate transfer of customer applications to it.

(12) In consultation with Systems, Communications and Operations sections, co-ordination of logistics to ensure that customers are able to access the services they require at the times they require.

(13) In consultation with the Operations Manager ensuring that both the Computing Service and customer security is protected.

(14) Arranging customer attendance at appropriate seminars, exhibitions, workshops, presentations etc. recommended by or provided by the Marketing and Sales Manager.

(15) Identifying customer training needs and liasing with the Customer Support Manager to meet these needs.

(16) Identifying any unsatisfied requirements of the customer for documentation or publications that the Computing Service is required to provide and rectifying the problem.

(17) Monitoring the duration of the contract or SLA and ensuring that all possible efforts are made to ensure continuation of the Computing Service to the customer.

(18) Regular reports to the Operations Manager and early advice of the cessation of, or renewal of, any contract or SLA.

(19) Advising the Marketing and Sales Manager of any possible new areas of business and participating in the planning of Customer Account Development.

Annex Four: Monitoring Tools - Web Addresses

Information on performance monitoring tools may be obtained from the supplier Web sites listed below.

http://adlex.th.net/services/asp_services.htm
www.bmc.com
www.brix-net.de/brix/index.htm
www.citrix.com
www.compaq.com
www.cai.com/unicenter/
www.firehunter.com
www.infovista.com
www.integnet.com
www.microsoft.com/technet/sms/
www.reporter.netcool.com
www.networking.ibm.com/paf/pafprod.html (Netview)
www.openview.hp.com
www.proxima-tech.com
www.sun.com/solstice/em-products/network/sunnetmgr.html
www.qwestcybersolutions.com
www.aprisma.com (Cabletron Spectrum)
www.tng.iac.es/
www.tivoli.com/products/index/netview/
www.caj.co.jp/products/tng_swdelivery_option_j.htm
www.responsenetworks.com/copyright/copyright.htm (VeriServ ™)
www.compaq.com.cn/service/IT_manager.asp (VitaSuite)

5

THE DOWNSIDE RISK; ALTERNATIVES TO SERVICE LEVEL AGREEMENTS; THE SLA PAYOFF

5.1 SLAs: Reasons for Failure

Not every SLA implementation is a success story. The possibility of failure has to be acknowledged. If the climate is not right for SLAs it might be preferable to consider implementing an alternative method to at least partially achieve the same ends.

But why do SLA implementations fail? They fail because:

- the business has no clear mission or direction
- the SLAs have no clear aim or direction
- the business application has no identifiable or accountable 'owner' or 'sponsor'
- the owner can't/won't see the benefits (they may later if benefits are proved by other SLAs)
- service definitions are vague, do not cover the whole service to the end-user's desk, or are not aligned to SLA aims or measurement processes
- SLAs have not been tailored to meet business needs (the relevant SL objective may have been missed, e.g. by concentrating on response rather than on data freshness)
- implementation spade-work has not been done: the ESCRIT framework is not in place (the text of the SLA is not supported by an adequate infrastructure)
- SLAs do not specify clear targets
- SLAs specify impractical, vague or unachievable targets
- the SLA is too detailed, cumbersome, indigestible and difficult to monitor
- the SLA is too skimpy and does not contain enough service criteria to be taken seriously
- the targets which are specified omit key factors so that, although the targets are met, they are not comprehensive enough to deliver appropriate service quality
- the SLA has not been followed up by real service orientation: it merely pays lip service to the concept
- measurements are too comprehensive and SLA monitoring is swamped with detail
- monitoring has been inadequate
- monitoring is not perceived by the customer as being objective and impartial
- service level management has not sufficiently resourced with:
  - skills
  - staff (headcount)
  - finance
  - the Service Level Management Toolkit

- the SLA project lacks control:
  - management is insufficiently involved
  - problems are not effectively followed up
  - Service Review Meetings have withered
  - service has foundered on political in-fighting within the IT Department
- no sanctions are applied for missing targets
- no remedial action is taken when targets are missed.

To implement Service Level Agreements the whole of the IT area must speak with one voice and present one face to the customer.

5.2 Alternatives to SLAs

What alternatives to SLAs are there? Most alternatives fail to create a true meeting of minds between the customer and the computing service, do not promote service culture as effectively and do not align the computing service to business needs as directly as SLAs do. Nevertheless, if SLAs are not practicable within an organization, they may be worth considering. The main options are:

- Performance Indicators
- Availability and Response Targets, which may be:
  - implicit
  - explicit
  - standard
  - model
- Benchmark Checks
- Customer Satisfaction Surveys.

5.3 Performance Indicators

Performance indicators may be published and circulated to customers (possibly in a newsletter) or discussed at Customer Service Review Meetings.

Current performance may be compared with past performance, perhaps by means of a trend graph, so customers can see whether service is consistent, improving or declining. Performance indicators are implicitly a one-way contract: the customer may not be tied to volume or throughput forecasts or to meeting deadlines, yet implicitly the computing service is expected to deliver consistent service levels.

Performance indicators may cover:

- hardware faults
  - mainframe, midrange and client/server equipment
  - storage devices
  - networks
  - printers
  - peripherals

  MTBF, MTTF downtime
  by day/hour number of
  users impacted and cost of impact

- software faults impacting
  - mainframe, midrange and client server services
  - networks
  - printers
  - peripherals
  - distributed processors
  - applications
  - services
  - facilities

MTBF, MTTF downtime
by day/hour number of
users impacted and cost of impact

- total problems
  - number, severity and time to rectify
- response time
  - by appropriate transaction type
- job completion
  - time in various categories
- utilization and capacity
  - forecast vs. actual
- output handling
  - against target deadlines.

5.4 Availability and Response Targets

Implicit

A batch job may be processed to produce output of financial management information. Although no formal targets have been set, the computing service knows that the later the results are produced, the less their value. It therefore aims to produce the output by the third working day of each period. It depends on the Finance department to input data but even when data is input later than usual, the computing service makes every effort to produce the output on time. In this case, implicit targets have been established by custom and practice.

Similarly with response. Where no formal response targets have been set, a customer may phone the Help Desk to complain about "slow" response. If the computing service agrees that response is "slow," it implies a response target that is better.

Implicit targets are common as an organization embraces computing for the first time, while the service is immature. But implicit response targets are an abdication of responsibility and can only ultimately lead to dissatisfaction of both customer and computing service. They are no substitute for service management.

Explicit

Explicit targets are frequently one-sided. They may be designated by the computing service as "reasonable".

In the example below no formal response time criteria had been set in consultation with the users. However, internally the criteria at Figure 5.1 were used as 'informal objectives'. These objectives were being met, and the majority of users appeared satisfied with the performance of the system.

We frequently see SLAs with, say 95% response targets – without any statement as to what happens to the final 5%. These may be the most important 5%! It is important to close down such targets.

Figure 5.1. Explicit Service Targets

EXPLICIT SERVICE TARGETS

TSO	Period 1	-	Average response < 1 second		
	(prime outside peak)		- 80% transactions completed within 1 sec		
	Period.2	-	Average response < 3 seconds		
	(peak)	-	80% transactions completed within 4 sec		
Batch	Class S	-	80% completed within	2	mins.
	Class A	-	80% completed within	10	mins.
	Class B	-	80% completed within	40	mins.
	Class C	-	80% completed within	80	mins.
	Class D	-	80% completed within	160	mins.

100% targets are 80% targets + 50%

Alternatively, explicit requirements may be imposed by the customer on the computing service.

"Payslips will be produced for circulation by 0800 hours on the last Friday of each month. Salary Payroll Tapes will be sent to BACS to arrive by 1600 hours on the last Friday of each month" or *"85% of transactions type A shall be completed within 2 seconds."*

Explicit targets may be expressed in a system specification for one large application expected to run on already loaded corporate hardware and Operations may not have been a party to the specification process. clearly an untenable situation.

Standard

Standard targets are often used by bureaux. They may state the terms and conditions on which work is to be undertaken (usually with the implication that they are not negotiable, although in practice they may be). Standard targets may impose obligations on both customer and computing service. Again, payroll is a good example.

> *"The customer will provide completed input by 1200 hours on the last Wednesday of each month for a Friday pay-day."*

Model

Model targets are usually the Performance Indicators (and perhaps Explicit Service Level targets) set by the computing service but they are declared to the customer and represent a framework within which the customer and the computing service may negotiate. This negotiation may, for instance, involve improving response by changing job priority. at a consequently higher cost to the customer.

Equally the negotiation could result in the customer bringing forward the deadline for completion of output, or smoothing work over a longer period. Model targets are the closest to SLAs in that a dialogue involving negotiation is established between customer and computing service.

5.5 Benchmark Checks

The benchmark approach involves taking several typical or representative on-line transactions (a 'basket'), establishing response time service level targets for completion of them, and periodically checking that this has been achieved. Taken in conjunction with model availability targets and model or explicit batch job targets they can make a substantial contribution to service level management without a SLA or they may be included as part of a SLA.

Benchmark checks are particularly useful for "difficult to measure personal computing services like electronic mail and CAD systems.

For an office system the basket could include, say:

- 1 logon
- 10 reviews of mail items awaiting attention
- 5 creations of an item of mail
- 5 document retrievals of stored mail
- 3 diary checks
- 1 appointment booking.

For a CAD system, the benchmark could be a typical drawing or a representative diagram like Figure 5.2a. This is an example of a live benchmark 'basket' test. The CAD group suffered from poor but unmeasured response time and friction occurred between the group and the computing service, who could not identify that response was a problem. The situation was resolved by periodic running of the benchmark test at the CAD stations.

The first resultant report (Figure 5.2b) clearly identified that there was a problem, with 16 missed targets in under 4 weeks. The CAD system was running at specification limits in terms of line lengths and the CAD team and the computing service jointly began a program of system rationalisation and tuning. The next report, covering 6 weeks, showed an improvement to only 7 missed targets. After further tuning, only 3 missed targets were reported over 9 weeks.

5.6 Business Satisfaction Analysis

We have seen how a Customer Satisfaction Survey (following Chapter 2) can provide a base line from which to measure the customer's perception of the quality of the computing service. We have also seen how a Service Level Survey (following Chapter 4), completed by the customer with help from a representative of the IT Department, can provide detail on which a SLA can be based. Some organizations go further and use Business Satisfaction Analysis techniques instead of SLAs. They start with a formal business analysis process which relates both customer's objectives and the computing service's objectives to corporate goals.

Figure 5.2a: A CAD Benchmark

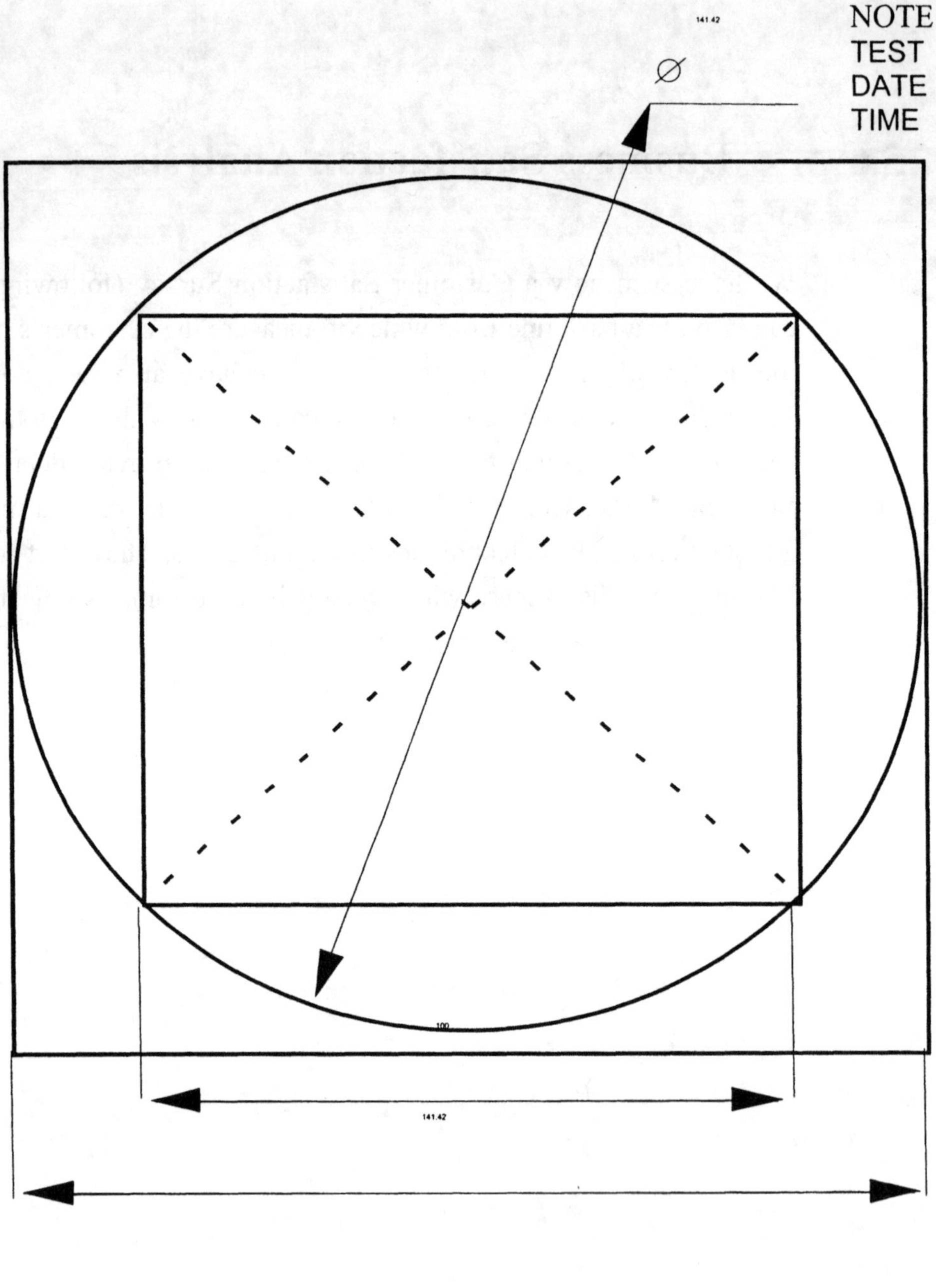

Figure 5.2b: CAD Response Benchmark Drawing Time

30th January - 24th February

DATE	TIME	BENCHMARK TIME	NO. OF JOBS
30.1	16.00	1.38	201
1.2	10.30	1.43	22
2.2	10.30	2.08*	337
	14.30	2.01*	326
3.2	09.00	2.02*	274
	12.30	1.43	273
	14.30	2.18*	304
	16.10	2.10*	325
4.2	10.15	1.57	304
	11.30	2.04*	328
	13.00	1.52	266
5.2	10.00	1.54	324
	11.30	2.16*	344
10.2	09.00	2.05*	312
	12.00	1.58	329
	14.50	2.12*	301
	16.00	2.04*	318
11.2	09.00	1.48	294
	14.00	1.54	307
	16.00	2.01*	321
12.2	09.00	2.12*	254
	11.00	2.08*	296
	14.30	2.22*	355
16.2	16.00	1.53	339
17.2	12.00	1.50	300
23.2	11.00	1.58	321
	13.00	2.10*	315
24.2	10.45	1.55	297
	16.00	2.01*	326

* Denotes missed targets

Figure 5.2c CAD Response Benchmark Drawing Time

30th March - 13th May

DATE	TIME	BENCHMARK TIME	NO. OF JOBS	REMARKS
30.3	14.20	3.2 1*	340	40 seconds initial access to file
2.4	14.45 14.55	1.59 2.19*	328 336	system slow on drawing
6.4	14.05	2.14*	334	
7.4	10.05	1.53	246	
8.4	15.20	1.51	322	
10.4	15.50	1.49	210	
14.4	09.15	2.02*	294	
27.4	10.30	1.59	312	
28.4	14.45 15.35	1.52 *2.05**	328 334	
29.4	15.30	1.54	296	
30.4	09.30	2.24*	300	Slow response to dimensions & notes
1.5	11.15 15.30	2.10* 2.06*	329 313	Slow access to files. Slow dimensions & notes. System stops for seconds regardless of operation
	6.5	16.15	1.52	336
7.5	14.30	1.54	329	
8.5	15.45	1.51	328	Complaints of being slow. Not shown by this test?
11.5	11.00	1.50	331	
13.5	16.00	1.58	324	Slow access to files
*Denotes missed targets				

Figure 5.2d: CAD Response Benchmark Drawing Time

29th July - 3rd October

DATE	TIME	BENCHMARK DRAWING TIME	JOBS	REMARKS
29.7	10.45	1.51	294	
31.7	15.35	1.53	286	
4.8	2.30	1.48	281	
5.8	14.45	1.56	290	Slow access to files.
				Complaints of being slow.
7.8	15.35	1.58	201	Complaints of being slow.
				Not shown by this test.
10.8	14.30	1.50	-	
13.8	8.50	1.34	195	
18.8	14.50	2.04*	-	Continually slow on selection from screen
19.8	14.45	1.51	291	
24.8	9.30	1.52	284	
26.8	10.00	1.48	190	
28.8	10.35	1.49	271	
2.9	11.00	1.57	262	System slow. Not shown by this test.
7.9	10.30	1.54.	-	
10.9	10.20	2.33*	257	Slow on all operations. Have to select most options twice.
18.9	10.40	1.50	-	
29.9	10.30	1.52	263	
29.9	15.50	1.57	291	
7.10	14.30	1.48	302	Slow selection from screen.
14.10	9.35	1.53	149	Prior to this test the system was very slow and eventually stopped
20.10	12.30	1.31	198	
23.10	16.30	1.54	-	
27.10	11.00	1.50	-	.
2.10	14.45	2.23*	240	All screens lock up for 10 to 20 seconds
3.10	16.15	1.58	321	Slow on most operations

* Denotes missed deadlines

The Customer Satisfaction Survey may be used to identify the customer's perception of the value of the service.

This is followed by a full-scale review of all of the computing service's offerings to its customers. The computing service will have a range of products (e.g. hardware, software or documentation). It will also be providing a number of services (e.g. on-line, batch, specific databases, help desk, information centre). It will be operating various procedures (perhaps for applications change management, for ordering terminals or personal computers, for requests for new services). The computing service will operate a number of processes (examples include production of hardcopy or microfiche output, database updating). It will also run a number of systems (that is, operating systems supporting applications run on specific hardware with identifiable peripherals).

All these products, services, procedures, processes and systems should be evaluated against the actual needs of the customers in fulfilling their contribution to corporate objectives.

Information analysis techniques may then be used to identify the relevance, timeliness and value of all computing service activities in contributing to the achievement of corporate plans.

In the mature computing service many activities result from history rather than need. Many services are provided in a particular way or to a particular format because that was the easiest way of meeting what was expressed as a customer need at the time. The analysis process gives the opportunity to identify what the present need really is and how well it is met by present services.

The Business Satisfaction Analysis may be repeated annually to ensure that the computing service maintains appropriate quality and relevance to the business.

This type of Business Satisfaction Analysis represents a thorough audit of the value of all the work of the computing service. It should identify all those activities which are of little value or relevance and help to adjust priorities so that resources can be allocated to key areas. This approach may in fact achieve more than SLAs taken in isolation since it provides an opportunity to rank all computing service activities in order of importance.

The benefits of Business Satisfaction Analysis are that:

- it aligns the computing service to business needs
- it leads to better utilization of computing assets
- it emphasises the value of what is produced rather than the mechanics of producing it
- it improves the effectiveness of the computing service by identifying non-essential activities
- it increases the customer orientation of the computing service
- it increases customer satisfaction
- it monitors customer satisfaction.

The Service Management Toolkit is as important in this scenario as for SLAs, and the analysis has to have sufficient management support to ensure that any recommendations for change are followed through. The recommendations may involve changes in either the customers' area or the computing service's.

Whereas the SLA is an excellent vehicle where there is substantial departmental autonomy or in a de-centralised environment, the Business Satisfaction Analysis can be very appropriate for an organization with a strong centralist culture.

5.7 The SLA Payoff: A Success Story

Although reasons for failure have been examined it is encouraging to finish on a success story.

A small American bank looked at its capacity plan and realised that unless something was done, a costly IBM mainframe upgrade would be needed. At that time the bank had performance targets decided by its IT Department in addition to SLAs. The IT objective was to excel SLA targets, since when the existing mainframe had been installed there had been excess capacity. Customers had thus got used to excellent response times which had become a de facto standard.

By going back to the SLA requirements and comparing present

performance with them, the IT Department discovered that the response time for most transactions and most batch input queue times could have been doubled and still remained with the SLAs.

Utilization had increased from 67% of available capacity to 82%. The critical peak period was the approaching year-end: getting over that hump was crucial. Performance monitors identified the bottleneck to be in the data storage area. This was resolved by installing cache storage directors, upgrading to newer, more efficient disks and disk tuning. Performance was monitored daily to ensure SLA targets were met. Some performance enhancement software products were also bought.

The result of all these measures was to defer purchase of a new mainframe showing a net saving of $1.52M.

Savings could be higher if the only other option to extending the life of existing hardware is to buy new hardware close to the end of the product life cycle, when rapid depreciation can take place as soon as the new model is announced.

This case study has several lessons. It highlights:

- the benefits that can be gained from intelligent application of SLAs
- the dangers of over-providing service quality
- the inter-relationship between SLAs, capacity and performance management
- the need for installation and service management methodologies and tools to be in place to support SLAs.

5.8 Where Next?

The need for a business-oriented utility computing service will inevitably lead to automation of many of the service management functions. In telecommunications, manual switching gave way to mechanical and subsequently electronic switching and we are now in the era of dynamic network management using intelligent software. In exactly the same way computing service functions will inevitably become automated to reduce human error and increase cost-effective quality and consistency of service. Some of the present generation of automated system and network monitoring and management tools extend to include:

- automatic rectification of most
  - hardware error conditions
  - software error conditions
- environmental monitoring and management
- automated recovery from system failure
- links to disaster recovery and continuity planning software providing automatic call-out or even automatic invocation of stand-by arrangements
- automation production of SLA system performance reports.

5.9 Conclusion

There are alternatives to SLAs, but only SLAs get the message across that the computing service is embracing the full service culture.

Integrating IT with Corporate Objectives has been identified as the main concern by 47% of IT managers. SLAs can help to align the IT function with business needs, but they cannot be effectively implemented without the Service Management Toolkit being in place; without adequate resource and without commitment from all involved. And once involved, they continue to need servicing.

But for how long? This book started by saying Information Technology was becoming just another utility. But we don't usually have specific Service Level Agreements with utilities: for instance, there is a general implicit service level of 100% availability. If this is the case, perhaps SLAs are just an interim stage, until so much resilience has been built into our systems that they become as robust as other utilities. This is not a long way off for most enterprises and is a reality for some now. But there will always be newer technologies coming along and new commercial services being offered. We will still need the Service Level Toolkit for internal management of the service. It is over ten years since the first edition of this book was published and SLAs are as valid now as they have ever been.

APPENDICES

APPENDIX A

SERVICE LEVEL AGREEMENT CHECKLIST

This checklist is intended to assist in the development of a Service Level Agreement. Since Service Level Agreements can be used to describe a variety of services and parties, the elements appropriate to any specific SLA will depend on the circumstances.

The Checklist, though long, is not exhaustive. A SLA including all these items would be cumbersome and probably impractical. It is not intended that all items should be included in a SLA, only that all items should be considered. Some of them may be reflected in job descriptions; in Management by Objectives targets; in relationships between customer support staff and customer; or between workload forecasting and capacity planning.

Grateful acknowledgements are due to George W. (Bill) Miller, American Airlines, (see Bibliography) from whose work this Checklist has been developed over the years. Any aberrations are the Author's, not Bill Miller's!

SERVICE LEVEL AGREEMENT CHECKLIST

Purpose of Agreement

Date of Effect

Duration of Agreement

Parties to Agreement and Contacts

Service Manager
Site/Accommodation Manager
IT Customer Account Manager
Customer
Customer Sites/Accommodation Manager
Customer contacts.

Description of Service

Overview and Objectives
Site(s) to which Service is to be provided
Definition of Service:

- Definition of Service regimes
  - prime time
  - non-prime time
  - weekend/public holidays

- Supercomputing services
  - specify and describe each application
  - including storage requirements
    - . primary
    - . secondary
    - . archive

- real-time
- interactive work
- batch jobs

- Batch Services
    - specify and describe each application
    - including storage requirements
        . primary
        . secondary
        . archive
    - and output requirements

- Real-time services - interactive work
    - specify and describe each application
    - including storage requirements
        . primary
        . secondary
        . archive

- Telecommunications services
    - voice
    - data
    - terminal provision
    - terminal maintenance
    - network
        . WAN
        . LAN
    - private circuits
    - network components
        . bridges, routers, controllers
        . modems
    - front-end processors
    - intranet
    - extranet
    - internet

- Outsourcing & Facilities Management Services
    - hardware covered
    - operating systems covered

- applications covered
- telecommunications and network
- maintenance
- consumables
- environment
- security

- Applications Development Services
  - business analysis
    - specification
    - system analysis
    - system design
    - testing
    - implementation

- Distributed Processing and Desktop Services
  - hardware provision
  - hardware installation
  - hardware maintenance
  - software provision
  - software maintenance
  - configuration management and version control
  - anti-virus and FAST activity
  - application development
  - security, back-up

- Technical Support Services
  - operations
  - system support
  - quality control
  - database management
  - data management
  - capacity management
  - security management
  - co-ordination of in-house suppliers
  - co-ordination of external suppliers
  - maintenance management
  - advice on hardware selection
  - advice on software selection

- other consultancy services

- Problem Management
  - services covered
  - Help Desk (Level 1)
  - second and higher level of support
  - problem resolution objectives
  - escalation procedure

- Education and Training Services
  - IT
  - customer
  - in-house
  - external

- Administrative Services
  - accounting and charging
  - leasing arrangements
  - equipment rental
  - software rental/royalties
  - organization and reporting lines
  - insurance/risk management
  - contract management.

Workload Forecasts

Supercomputing:

- as batch and interactive below
  - plus memory requirements
    . peak
    . average
    . over each regime

- forecast growth peak and average over each regime over 6 months, 1 year, 2 years

Batch Jobs

- daily volumes
  - peak
  - average
  - over each regime

- weekly/monthly volumes
  - maximum
  - average
  - over each regime

- forecast growth, peak and average over each regime over 6 months, 1 year and 2 years

- workload above which Service Level will not be guaranteed

Real-time Services - Interactive Work

- hourly transactions volumes
  - peak
  - average
  - over each regime

- daily peaking

- over each regime

- total number of users

- total number of simultaneous users

- forecast growth, peak and average over each regime over 6 months, 1 year, 2 years

- workload above which Service Level will not be guaranteed

Call Center

- identify products covered
- existing workload
- average duration of inbound calls
- average duration of outbound calls
- average number, per day / hour, of inbound calls
- average number, per day / hour, of outbound calls
- hourly volume - peak, average, over each regime
- agent productivity standards
- number of calls hitting "engaged"
- average call abandonment time
- call abandonment rate
- prioritization procedure
- escalation procedure
- new products scheduled to move from development to production
- operating system upgrades scheduled
- hardware changes scheduled
- middleware changes scheduled
- business application changes scheduled

- network changes scheduled
- environmental changes scheduled
- forecast growth, peak and average over each regime over 6 months, 1 year, 2 years

Telecommunications and Data Communications Workload

- for each network service (e.g. ATM, Frame Relay, intranet etc)
- hourly volumes
    - peak
    - average
    - over each regime
- daily volumes
    - peak
    - average
    - over each regime
- total number of users
- number of simultaneous users
- forecast growth, peak and average over each regime over 6 months, 1 year, 2 years

Outsourcing & Facilities Management

- service hours
- Service Level requirement (by use of other relevant sections of this Checklist)

- forecast growth, peak and average over each regime over 6 months, 1 year, 2 years

Applications Development Workload

- existing workload
- human resource available (permanent and contract staff)
- corporate plan (systems implications)
- headcount restraints
- resource forecasting methodology (e.g. COCOMO)
- development tools (CASE, 4GL, 5GL etc)
- project management responsibilities (sponsor, client, executive)
- project management methodologies
- design standards
- design methodologies
- quality assurance and quality control standards
- testing responsibilities
- documentation responsibilities
- forecast growth, peak and average over each regime over 6 months, 1 year, 2 years
- list
    - applications priorities
    - resource requirements

- implementation deadlines
- project schedules

Distributed Processing and Desktop Services

- existing resource requirement
- daily peaks and average demand
- weekly peaks and average demand
- forecast growth, peak and average over each regime over 6 months, 1 year, 2 years

Technical Support Services

- for each support service specify
  - existing resource commitment
  - daily peaks and average demand
  - weekly peak and average demand
  - forecast growth, peak and average over each regime over 6 months, 1 year, 2 years

Problem Management

- identify support services covered
- existing workload
- hourly volume - peak, average, over each regime
- agent productivity standards
- number of Help Desk calls hitting "engaged"
- average call abandonment time
- call abandonment rate

- number of calls not logged
- average duration of inbound calls
- average duration of outbound calls
- average number, per day / hour, of inbound calls
- average number, per day / hour, of outbound calls
- prioritization procedure
- escalation procedure
- new applications scheduled to move from development to production
- operating system upgrades scheduled
- hardware changes scheduled
- network changes scheduled
- environmental changes scheduled
- forecast growth, peak and average over each regime over 6 months, 1 year, 2 years

<u>Education and Training</u>

- on-the-job training
- subscription TV training
- video training
- interactive video

- computer-based training
- intranet / internet based training
- “classroom” training in-house
- external courses
- prerequisite training
- tutor/student ratio
- hardware/student ratio
- optimum
- maximum
- frequency of training
- structural training schedule
- marking and certification
- forecast growth, peak and average (eg from new applications or new customers) over each regime over 6 months, 1 year, 2 years

<u>Administrative Services</u>

- for each service
- existing resource commitment
- forecast requirement daily peak and average; weekly peak and average; by all regimes covered

- forecast growth peak and average (e.g. from new applications or new customers) over each regime over 6 months, 1 year, 2 years.

Response Levels

Establish theoretical maximum throughput and hence from capacity and workload forecasts

Supercomputing Services

- use checklists below

Interactive Work

- response time for specified type of transaction or for baskets of transactions at
  - host processor or server
  - customer terminal

- percentage of transactions falling within specified response targets for each type of transaction

- reduced response times related to increased transaction workload

- response targets peak, average, for each regime

Batch jobs

- turnaround - time elapsed between

  - job start to job end
  - receipt of input to delivery of output
  - average times compared to targets and related to volume

- compliance with schedules
- number of deadlines missed

Call Center

- time limits for "on hold"
- time limits for agent resolving each call on the phone
- time limits for issue resolution
- call backlog
- system response time

Telecommunications and data communications Services

- network response times
- percentage of transaction falling within specified response
- reduced response times related to targets for each type of transaction workload
- response targets peak, average, for each regime

Outsourcing & Facilities Management

- compliance with service hours
- standard service response (by use of relevant sections of Checklist)
- response to ad hoc requirements

Applications Development Services

- delivery from development to production compared to project planned timescale

- backlog

Distributed Processing and Desktop Services

- as for relevant sections of Checklist for Batch and Real-time

Interactive work

- compliance with "hours of business"

- elapsed time from service request to delivery of service

Technical Support Services

- defined for each service

- compliance with deadlines

- service frequency compared to schedule

- reaction time for new requests

- reaction time for urgent requests

- backlog

Problem Management

- time limits for “on hold”

- time to respond

- time to provide work round
- time to fix
- escalation invocation
- backlog
- system response time

Education and Training

- compliance with published training schedules
- compliance with individual course timetable
- time to initiate new courses
- time to revise existing courses to reflect changes
- backlog

Administrative Services

- elapsed time to register new users
- elapsed time to give new users access to services required
- timeliness of provision of accounting information
- elapsed time between customer request and completion of service

provision

- backlog

Accuracy

- Accuracy targets would be aimed towards a zero defect service

<u>Batch Jobs</u>

- accuracy of input (ratio of errors to volume)
- definition customer/IT accuracy responsibilities
- conditions for re-run
- number of re-runs
- impact on response level if customer quality
- targets are not met
- meantime between operator error during each regime
- number of errors during each regime
- meantime between systems error during each regime
- number of systems error during each regime
- number of patches or bug fixes applied (by time)
- number of changes applied (by time)
- ratio of tape read/write errors
- frequency of misdirected output
- frequency of consumable "out of stock"

Interactive Work

- data transfer error rate (ratio to volume)
- number of and frequency of outages from same cause

Call Center

- correctness of information
- completeness of information
- number of customer repeat calls on same issue / product
- number of complaints of poor information

Telecommunication and Data Communication Services

- incidence of outage of circuits
- incidence of outage of primary routes
- data transmission error rate (ratio to traffic)
- frequency and number of misroutes
- package loss rates
- number and frequency of outages from same cause

Outsourcing & Facilities Management

- mean time between operator error during each regime

- number of errors during each regime
- mean time between systems error during each regime
- number of systems error during each regime
- number of patches or bug fixes applied (by time)
- number of changes applied (by time)
- select from other sections of Accuracy Checklist as appropriate

Applications Development Service

- mean time between errors
  - at specification
  - on testing
  - in production
  - per lines of code
- number of errors
  - at specification
  - in testing
  - in production
  - per lines of code
  - per 1000 Function Points
  - impact of errors
- number of bug fixes applied (by time)
- number of requests for changes (by time)
- number of changes applied (by time)

Distributed Processing and Desktop Services

- select relevant items from Accuracy Checklist section

Technical Support

- mean time between errors
- number of errors
- impact of errors
- correctness of information
- number of revisions of published information
- number of reiterations
- number and frequency of complaints
- number of Help Desk calls on identical problems

Problem Management

- correctness of information
- completeness of information
- number of customer repeat calls on same problem
- number of Help Desk calls on identical problems

Education and Training

- student assessment of course quality (course critique forms)

- performance on any tests (before/after)
- line management feed-back on effectiveness of training
- retraining level

Administrative Services

- frequency and number of errors
- number and frequency of iterations
- number and frequency of complaints.

Availability

Production Services (supercomputing, real-time interactive and batch). For each service:

- define "business hours" – i.e. scheduled times of availability
- define "availability" to take account of part-crippled service
- percentage scheduled availability at customer terminal for each regime and at peak period

Call Center

- define whether “availability” includes
  - use of answer phones, Interactive Voice Response (IVR), Fax in Fax Out (FIFO) or other unattended systems
- define "business hours" manned Call Center

- actual availability compared to “business hours”
- escalation procedure

Telecommunications and Data Communications Services

- as for Production Services

Outsourcing & Facilities Management

- as for Production Services

Applications Development Services

- resource delivered compared to resource required by project plan
- project duration

Distributed Processing and Desktop Services

- as for Production Services

Problem Management

- define whether "availability" includes
  - use of answer phones
  - use of voice response systems
  - customer’s use of expert systems
  - customer searching sub-sets of, or actual, problem database
- define "business hours" for
  - Help Desk

- second and higher level support

- actual availability compared to "business hours"
- escalation procedure

Technical Support

- define "business hours"
- define cover arrangements for leave, sickness, etc.

Education and Training

- define "business hours"
- course schedules and timetables
- location of courses

Administrative Services

- define "business hours"
- define cover arrangements for leave, sickness, etc.

Constraints

Dependency on:

- internal supplier with whom no SLA exists
- external supplier with whom contract does not explicitly cover the SLA requirements

- shared resources
- business priorities remaining constant
- customer priorities remaining constant
- capacity plans being approved
- staffing plans being approved
- budget being approved
- customer usage aligning with forecasts

Charging and Recovery

- Cost or Profit Centre?
- cost notification, cost allocation or charge-out?
- breakeven or standard costing?

Charging Definition

- Computer Resource Units
- Computer Workload Units
- Business Workload Units
- items to be included in charging
  - computer resources
  - overheads
  - people time
  - consumables

- technical support
- consultancy
- bought-in services
- profit?

Demand Management by Pricing

- higher price for "gold star" service (better quality)
- higher price for "express" service (quicker service)
- higher price for peak service
- higher price for higher workload (to discourage overload)
- discount for higher workload (to encourage full utilization)
- discount for non-peak time or for non-prime time regimes

Personnel Incentives

- contracts of employment
- definition of gross misconduct (i.e. firing offences)
- job descriptions
- performance related pay
- Management By Objectives
- Non-financial incentives

Monitoring of Performance against SLA

- Responsibilities of
  - IT
  - the customer
  - co-responsibilities

- Define
  - what is to be measured?
  - where it is to be measured?
  - what is the basis of measurement?
  - what source data is to be used?
  - what software tools are to be used
  - what procedures are to be adopted?
  - what assumptions are made?
  - what reports will be produced
    - . by what method
    - . in what format
    - . to whom
    - . how often.

- **SLA Review and Renegotiation Procedure**

Penalties and Rewards

Thresholds

Composition of Balanced Scorecard

Penalties

Bonuses

APPENDIX B:

Example Desktop Support Metrics

PC SUPPORT

One very important aspect of designing SLAs is defining service products and specifications. The following is intended to act as a to providing in-depth PC support. If this is a chargeable service, the cost needs also to be stated.

Services	Service Level Issues & Metrics
Network	
Cabling	lead time; deadlines; defects out-of-stock frequency;
Diagnostics	lead time; deadlines; effectiveness
Tuning	lead time; deadlines; network efficiency; tuning effectiveness results)
Installation	
Cabling	lead time; deadlines; defects; out-of-stock frequency
Delivery	lead time; deadlines; out-of-stock frequency; damage; location
Set-up	completeness; functionality; integration - hardware - software - network
Data transfer	lead time; integrity; completeness; accuracy; deadlines
Customer Support/ Problem Solving	
Self-diagnosis cards	lead-time; out-of-stock frequency; ease of use; comprehensiveness; accuracy

Services	Service Level Issues & Metrics
Help Desk	'opening hours'; number of problems resolved on the telephone within 5 minutes; no of problems resolved within defined deadlines within each priority; number of call-backs on same problem; complaints level; escalation procedure (including out-of-hours)
Consultancy	lead time; deadlines; effectiveness; deliverables
Disaster Recovery	
Loan equipment	availability (out-of-stock frequency); response time; inventory of hardware and software available on loan
Facilities	size; standard of fit-out and availability of accommodation; lead time and response time; duration of occupancy; network access; simultaneous invocation - what happens
Salvage	clean-up services available; lead time; clean-up time; collect or customer deliver; warranties; subsequent maintenance implications; maximum volume and throughput

Services	Service Level Issues & Metrics
Back-up	when done; who is responsible; nature (full image, incremental; systems and data or data only); where stored (on or off-site); fire storage standards (e.g. VDMA Specification 24991 RA RG626/7 S120 DIS - a 2-hour fire protection rating); arrangements for withdrawal of back-up; customer/support responsibilities for integrity of back-up and recoverability
Virus scanning	customer responsibilities; standards; turn-round time; maximum volumes
Virus clean-up	customer responsibilities; turn-round; limitations; hit rate;
Consultancy	lead time; deadlines; effectiveness; deliverables
Maintenance	
Parts/labour	lead time; turnaround; defect rate; warranties; items covered
Remedial, hardware	lead time; response time; fix time; out-of-hours conditions; defect rate; warranties; items covered
On-site	hours of presence; skill level; number of staff; out-of-hours arrangements; fix time; items covered
Return to base	lead time; turnaround; defect rate; warranties; items covered

Services	Service Level Issues & Metrics
Preventative	schedule; defect rate; items covered; man-hour commitment
Replacement units	conditions; items covered; lead time; fix time; turnaround
Training	
Workshops	frequency; maximum/minimum numbers; location; start time; finish time; topics
Basic Training	frequency; maximum/minimum numbers; location; start time; finish time; topics; depth of course; objectives; trainer's skill level; lead time required
Advanced Training	frequency; maximum/minimum numbers; location; start time; finish time; topics; depth of course; objectives; trainer's skill level; lead time required
Continuation Training	frequency; maximum/minimum numbers; location; start time; finish time; topics; depth of course; objectives; trainer's skill level; lead time required
Network Management	see Network service above
Asset Management	
Acquisition	- hardware - software - maintenance - consumables - financing - lead time; deadlines; contract

Services	Service Level Issues & Metrics
	provisions; 'out-of-stock' frequency; discounts; customer responsibilities; procedures
Asset Register	- hardware - software
	accuracy; how up-to-date; notification responsibilities; maintenance procedures; level of detail (e.g. model number, serial number, version level)
Version Control	conditions of release of new version; lead time; frequency; collection of old versions; arrangements in respect of related documentation; related training issues (see Training above)
FAST Audit	(check for pirate software) customer responsibilities; lead time (if on request); frequency
Virus Scanning	see Disaster Recovery above

APPENDIX C:

Traditional, IT Oriented SLA

SERVICE LEVEL AGREEMENT

agreed between

COMPUTER OPERATIONS

and

____

For the period ________________

to________________ inclusive

STRUCTURE OF DOCUMENT

1. Purpose of Service Levels
2. System Description
3. Batch Work Service Level
4. Online Service Level
5. Peak Period Details
6. Computer Charging
7. Security
8. Impact of Loss of Service
9. Output Requirements
10. Change Control
11. Customer Support & Help Desks
12. Service Level Reporting
13. Contacts

Draft 1

SERVICE LEVEL AGREEMENT (SLA)

1. WHAT WILL THIS DOCUMENT ACHIEVE

The purpose of this document is to define the services to be provided by Computer Operations to its Customer Department.

The methodology is as follows:-

- Establish business requirements and define responsibilities.
- Identify suitable performance measures.
- Develop reporting procedures which identify effectiveness by function.

This will enable us to highlight trends sufficiently early to protect and improve service levels.

The objective of Operations is to maintain adequate capacity and resource to deliver the agreed performance targets for required volumes. To achieve this aim, the Customer must accurately determine any factors they are responsible for which can ultimately affect the deadlines.

Services will be made available at unscheduled times with the agreement of the Operations Controller. Extensions to services cannot be assumed as they may impact onto other systems.

AGREED BETWEEN:

CUSTOMER ______________________________ (Sign)

OPERATIONS ______________________________ (Sign)

There will be an interim review in 3 months. Unless changes are required this SLA is expected to stand between the periods of

xx/xx/xxxx and xx/xx/xxxx.

Formal liaison meetings will be arranged between Operations and the Customer to regularly review current service at appropriate intervals.

2. SYSTEM DESCRIPTION

e.g., Sales and Stock Reporting. This system provides Head Office and Stores Sales Information at way level. It also provides Stores with advice on the appropriate percentage of space in which to display each item within the Department.

2.1 Customer Department

e.g., The team reside in Head Office and are called Planning Development and Control. They maintain close links with all other Product 1 and Product 2 Departments to ensure planning co-ordination.

2.2 System Profile

e.g., This system and its deadlines are extremely critical. It is the tangible result of the weekend processing. The reports are intended to replace checking lists and the Sales Information is vital to the Buyer departments for planning purposes. More importantly the Board rely on these reports for their weekly figures.

2.3 Hardware Configuration Used

e.g. This service normally runs on the server cluster at Hayes.

2.4 Service Requirements

e.g. The SSR Database must be loaded between 0800 and 1800 hours. The reports must be at Head Office by 1400 hours each day.

2.5 Service Monitoring

e.g., The availability of the Database will be monitored along with the daily batch deadlines.

2.6 Special Conditions

e.g., The SLA may need to be re-negotiated should other services begin to share the same machine or volumes exceed expectation.

2.7 Measurable Objectives

e.g., The services are measured to ensure the continued growth of this System and the required deadlines can be maintained.

3. BATCH WORK

3.1 Services to be Provided

The following batch services will be provided by Operations and delivered where applicable to the Customer location to agreed timescales. Major service elements have been identified to simplify monitoring and improve reporting.

Service Elements	Frequency of Run	Day(s) of Run	Delivery time to department	Reliability Target
e.g., A. Runs	1 per week	Sun+Tue+Wed	0900	99%
B.				
C.				
D.				

The reliability of the Service is the basis for Service assessment. Where possible calculations will be performed on the computers at the I.T. Centre. For those that need to be determined manually, the following calculation should be used:-

Number of Deadlines
------------------------------------ x 100
Total Number of Deadlines

3.2 Customer Commitments

To meet the relevant deadlines, Operations are dependent on the Customer ensuring that the necessary data/forms are at the correct location and time for collection. Volumes are also required.

e.g.
- Data for midweek A runs to the bureau by 1500
- Data for midweek N runs to the bureau by 1800
- Data for weekend A runs to the bureau by 1800 Friday
- Capacity information should be within the range of plus or minus 10% for peak volumes

3.3 Volumes

Details of volumes are required to ensure resources are available to meet the service demand and avoid adversely impacting on other systems.

e.g.,
- Number of main file records
- Pages of printout
- Storage/recall requirements
- Expected growth in next 12 months
- Peak period dates and volumes

4. ONLINE SERVICES

4.1 Services to be Provided

Detailed here are the times at which the online services are required. Capacity information is again needed in order to meet requirements e.g. response time.

Service	Start	End	Mean Time Between Fail	Mean Time to Repair	Guaranteed Fix Time	Availability Target

Planned unavailability should not be included in expected availability total.

The achieved availability percentage should be calculated as follows: -

System availability
------------------------------- x 100
Expected Availability

4.2 Customer Commitments

To attain the Service Level requirements, Operations are dependent on the Customer supplying the following: -

- Capacity estimates, which must be within the range plus or minus 10%

- The maximum transaction type enquiry per minute which must not be exceeded.

- Number of database records which will not exceed ....

4.3 Capacity Information

The response time of an online system is the main feature by which the Customer assesses the system performance. Behind the scenes activity is taken for granted and irrelevant to a SLA. Accurate predictions depend on correct volume statistics provided by all Customers on a shared machine or network.

Peak response information also needs to be identified where perhaps the need is greater and yet the response can be slower. It is not satisfactory to purely consider peak information because the Customer becomes accustomed to the normal (non-peak, better) response.

4.4 Response Requirement

All response monitoring is performed at the I.T. Centre. It is normal in the industry to target 95% of the actions because complicated processing often leads to 5% variance.

95% TARGETS

Transaction Type	Peak Hours Volume	Average Hour Response	Peak Hour Response	Peak Hour Period
e.g. Updates	100 per min	1.5 sec	2.0 sec	1030-1130
Insertions	25 per min	0.8 sec	1.0 sec	1530-1630

The peak periods will be between the following periods.

xx/xx/xxxx to xx/xx/xxxx
xx/xx/xxxx to xx/xx/xxxx /

4.5 Equipment Reliability

It is the intention of the I.T. group to provide the relevant hardware to meet the required reliability targets. Hardware is subject to occasional failure and consequently the targets reflect achievable aims.

Equipment Type	Number of Units	Customer Availability Target	Contracted Callout	Contracted Fix Time
e.g., Private lines	3	99%	2 hour response	4 hours
PC Pentium	40	98%	4 hour response	6 hours

If callout times for local equipment are exceeded, please inform the Help Desk who will escalate accordingly.

5. PEAK PERIOD DETAILS

At Peak Periods the schedules will normally alter. For the relevant duration, Peak schedules will be agreed between the Customer and Operations Control.

To enable sufficient planning time, these schedules must be forwarded at least 4 weeks in advance.

The SLA cannot be used to measure performance on committed deadlines for the normal SLA period. This is because other systems on the same machine may also alter and consequently affect the service.

The Service Level Report, however, (Section 11) can be used to determine service satisfaction ratings.

6. COSTING THE SERVICE

To enable more effective Management control, a business charging unit will be defined and reported against by Business Accounting Period (4/5 weeks).

Reports will be supplied automatically by the computerised costing system and will be issued to the Customer Manager and I.T. Group Financial Control.

Discount arrangements: ______________________________

7. SECURITY

Security aspects include backup, handling of sensitive or security classified material and contingency plans

Backup

It is imperative that backup of data is completed to agreed timescales. Data may be irrevocably lost or severe delays to service could be experienced.

e.g.,

Files to be secured	Frequency	Scheduled Run Time
G.S.D. Database	once per day	0400-0700

Who will be responsible: ______________________________

Handling of Sensitive Information

e.g.,

Application ABC has password control by file, field and process

- all output from application CDE must be headed "Restricted Circulation," copy numbered, distributed by secure means and handed over to an authorised signatory against signatures.

Contingency Plans

e.g.

If Server A is not operational for more than 12 hours, application FGE is to be run at Standby location B.

8. IMPACT OF LOSS OF SERVICE

Actions required in the event of a disaster are considered in disaster recovery procedures and are outside of the scope of a normal SLA.

This section is required to identify the impact on the Business when a service is unavailable. This will help to cost a contingency service and avoid unnecessary high cost of hot standby facilities. The need for a standby machine is dependent on the priority of the system and the reliability of the service.

e.g.,

A. Downtime of 30 minutes per day is recoverable.

B. Downtime costing of 40 employees at £15 per hour = £600 per hour: for the Customer Department

C. Application GHI handles customer accounts with a daily order value of £25M. An outage of 8 hours could result in deferred cash flow to that value.

D. Late delivery of printout has no impact providing it does not exceed 1 hour Monday-Thursday and 30 minutes on Fridays.

E. Machine downtime may delay deadlines by an equal amount to the downtime.

F. If the service is unavailable for greater than four hours, a standby service is required.

9. OUTPUT REQUIREMENTS

This section will cover deadlines for printouts, microfiche, interface tapes etc.

These functions require more human intervention than other aspects for the computer runs and therefore are more prone to error.

Service Level targets should be set accordingly giving a higher target to 'remote viewing' print files due to the reduced levels for human intervention.

Output Name	Job Producing Output	Deadline to Customer	Target %
e.g., Priority Output			
e.g., Other Output			

10. CHANGE CONTROL

10.1 Objectives

- Optimise the rate of change while containing the risk of adverse impact on Customer Service Levels. Thus normally new versions of bought-in software will not be implemented within 6 months of release but will normally be implemented within 12 months of release.

- Monitor change installation to see objectives have been achieved.

10.2 Role

- Provide a central point to co-ordinate all computer change plans. (Hardware, Software, Network, Environment).

- Agree change plans with all interested parties, including customers.

- Report on effectiveness of implemented changes.

10.3 Service Maintenance Changes

- To maintain the mainframe and application services and to implement additional facilities, a number of changes will require Service availability. These changes are detailed below with agreed time windows for their implementation.

- The changes detailed below will not be counted against the achieved Service Level, as long as they are completed in the agreed time window.

Change Description	Reasons for Change	Time Window
e.g., To update the online service	To include increased performance facilities	1200-1300 Wed-Friday

11. SERVICE LEVEL REPORT

This document should be used as an aid to monitoring the service provided to the Customer by I.T. The rating column enables easy usage and common understanding. Unless otherwise stated all reports are on a rolling 4-week basis. Reports will be produced every 4 weeks.

Different systems can be so varied that it is more practical to rate the service by type e.g. Batch deadlines, Online deadlines, Supplier deadlines, Store deadlines. These are directly related to the major elements in both the batch and online services. Being type specific can therefore make scoring more meaningful and more accurately depicts potential problem areas. Where more than one service element is defined the average will be used for the overall system assessment.

This service will be reviewed via this report monthly.

<u>Service Rating Explanation</u>

10. Excellent
9. Good
8. Satisfactory
6. Minor Production impact causing inconvenience
4. Major Production impact, which is unacceptable
2. Very bad service
0. No service at all

e.g.,

Service Type	Customer Rating	Target	Weight Factor	Comments
Batch Work	10	9	50%	
Online Availability	6	9	20%	Not always available at start of day
Online Response	8	8	20%	Seems to be getting worse
Tradanet	4	9	10%	Suppliers received old information for system

<u>Service Rating</u>

The Service Rating can be calculated by taking into account the weighting factors when considered against targets.

<u>Footnote</u>

Operations will maintain details statistical information of all systems to objectively consider the cause of a low rating, aiming to improve service.

ANNEX A: CONFIGURATION MANAGEMENT

A.1 Document Approval

Approver	Sections Approved	Signature
	ALL	

A.2 Document Configuration

Title: Model SLA #4
Reference: TBD
Author: Andrew Hiles
Issue
Date
Location of Electronic Copy: TBD
Location of Paper Copy:
Owner:
Change Authority:
Distribution:

A.3 Change Log

Issue:	Date:	Details of Change:	Author

APPENDIX D: Example Simple Development SLA

MODEL SERVICE STANDARD

agreed between

DEVELOPMENT (BSS)

and

____

For the period ________________

to________________ inclusive

Section 1 Basis of Agreement

Core Services

Section 2

Other Non-core Services

Appendix 1 Statement of Service Charges including penalties and payment terms

Appendix 2 Parties to the agreement (signatories) including contacts and arbitrator

Appendix 3 Performance measures and statistics

1. BASIS OF AGREEMENT

1.1 Introduction

BSS operates as a trading agency within ABC Co to provide a range of services at prices that are competitive in the market place.

1.2 Purpose

The purpose of this agreement is to provide a sound framework for the provision of software maintenance and development services to the highest professional standards in support of the customer's business missions and objectives. It is designed to encourage a joint open partnership approach with regular and free exchange of information between both parties.

This agreement describes in sections **TBD** the individual services making up the total package of services provided by BSS and the reciprocal commitments made by the customer.

The services to the client and reciprocal commitments are shown in Section 2. Appendix 1 shows Charges for services, invoicing and payment arrangements.

The client's locations, units and operations included in this agreement are shown in Appendix 2.

1.3 Period of Agreement & Termination

This agreement will take effect from **(TBD)** and run for **(TBD)** years on a rolling contract. It is renegotiable within the last three months of each 12 month period. Either party may serve notice of termination to the other party to take effect from the anniversary of the start date or other agreed date if earlier providing a minimum of 6 months notice is given and such termination is agreed by the Executive.

1.4 Variation

The agreement will require renegotiation if there is a significant and sustained change in the scope of the original agreement by either party. Neither party will be liable for delay or failure to perform their part of the agreement due to circumstances beyond their control.

1.5 Communication and Access

The Contact Officers for this agreement in respect of the services described are detailed in Appendix 2.

1.6 Confidentiality

Both parties will comply with the terms of current legislation with regard to information in respect of its confidentiality, respecting individual client interests and in particular comply with ABC Co's policy on confidentiality of information.

1.7 Quality Assurance

BSS attaches the highest importance to the quality of service to be provided and will work with the client to determine and agree appropriate measures of quality.

1.8 Monitoring

To measure the effectiveness of BSS services, BSS will report to the client the actual performance measurements against the agreed targets as shown in Appendix 3. These reports will be at intervals agreed with the client and include any future performance measures agreed by both parties. Failure to meet target performance levels can (unless caused by the non-performance of any client commitments defined in the Service Specification) result in penalties as shown in Appendix 1 as agreed by both parties. In cases of dispute, either party may commission an independent audit of service level compliance.

1.9 Services outside Scope of Agreement

BSS shall be entitled to charge the client for services provided outside or over and above those stated in the agreement as agreed with the client.

1.10 Unforeseen Circumstances

If events occur which could not reasonably have been foreseen and which affect the substance of the agreement then either party may give not less than 6 months notice of their intention to end the agreement before completion of the agreed period. BSS will attempt to limit any adverse effect to clients due to unforeseen circumstances placed upon it.

1.11 Escalation and Arbitration

Every attempt will be made to rectify problems between the respective parties. However, in the event of failure of the two parties to agree, they shall refer the matter jointly to the appropriate arbitrator as agreed by both parties (see Appendix 2).

1.12 Benchmarking

In addition to on-going evaluation, benchmarking as agreed with the client will be undertaken annually in order to assess the effectiveness of the total service. Some limited external competition for the supply of the service can be undertaken but only as agreed by both parties.

1.13 ABC Co Policy and Statutory/ Regulatory Requirements

In the provision of the services contained in this agreement, both parties will ensure that ABC Co policy is implemented and that the necessary checks will be put in place to ensure compliance including statutory/regulatory compliance.

SECTION 2 SERVICE

AIM OF SERVICE

1. Introduction

1.1 Overview

BSS aims to deliver cost-effective, quality, pragmatic and responsive application services to its customers.

2. Services

2.1 Provision of Services

BSS shall provide the following services to the customer:

2.1.1 Software Development

The design, development, supply, delivery, installation, testing and implementation of new software programs and associated documentation as agreed with the customer.

2.1.2 Consequential Amendments to Software

Where alterations or amendments are made by BSS to hardware, software or network which have an adverse effect on Software, the provision and implementation of such amendments to Software as are necessary to remedy such adverse effects.

2.1.3 Enhancements to Existing Software

The design, development, supply, delivery, installation, testing and implementation of enhancements to existing software programs and associated documentation as agreed with the customer.

2.1.4 Preventative and Corrective Maintenance

Carrying out such preventative and/or corrective maintenance as shall be required by the customer from time to time. Such work will be controlled by issue in Releases.

3. Processes and Procedures

3.1 In delivering the above services, BSS will follow processes and procedures:

3.1.1 Project Management

Should the customer require a Project from BSS, the customer will provide BSS with a user requirement specification describing its requirements for the Project (the Specification) and other associated documentation in accordance with the BSS Project Manual **TBD**.

3.1.2 The Specification will identify the location where the Project is to be conducted and the location where subsequently the Project is to be delivered and installed.

3.1.3 BSS will then provide an Estimate of Project Cost.

3.1.4 On acceptance of the Estimate of Project Cost by the Customer, BSS will draw up a Programme of Work (BSS Customer Plan) in accordance with BSS Project Manual **TBD.**

3.1.5 On acceptance of the Programme of Work by the customer, BSS will execute the Project in accordance with the Programme of Work to time and budget using best industry standards and structured methods.

3.1.6 BSS will undertake appropriate testing to ensure the Project meets its Specifications and Service Standards and submit the Project to Acceptance by the customer and Operations.

3.1.7 BSS will conduct a Customer Satisfaction Survey and Post Implementation Review within 6 months of Acceptance of the Project.

Service Standard	Customer Commitments
1. <u>Estimate:</u> Provided within **TBD - 10?** days of request 2. <u>BSS Customer Plan (Programme of Work):</u> Provided within **TBD** days of receipt of request for Project 3. <u>Development Cost:</u> Not more than £**TBD** per 100 Function points 4. <u>Development Cycle Time:</u> **TBD** days per 100 function points 5. <u>Programmer Productivity:</u> **TBD** Function Points per normal working month 6. <u>Defects:</u> **TBD** Bugs per 100 Function Points in 20 working days of Acceptance 7. <u>Analysis:</u> < **TBD** Enhancement Requests within 20 working days of Acceptance 8. <u>Availability of delivered Project:</u> **TBD**% availability between the hours of **TBD** 9. <u>Reliability of delivered Project:</u> No more than **TBD** service breaks in each rolling 20 days 10. <u>Maintenance cost:</u> Not more than $**TBD** per 100 Function Points	1. The Customer Specification will be comprehensive, covering all aspects identified in the BSS Project Manual **TBD** and accompanied by: a Cost/Benefit Analysis; a Business Risk Analysis; a Business Impact Analysis; all of which will be kept updated during the course of the Project. 2. Response to the BSS Customer Plan will be made in **TBD** days (exceptions may result in changes to the Project timescale contained in the Customer Business Plan). 3. Resource will be provided in the timescales and to the skill levels agreed. 4. Deadlines agreed for deliverables by the Customer will be met.

APPENDIX 1

STATEMENT OF CHARGES

Core Services

1. Service — Breakdown of cost (e.g., labour, materials, overheads) — Penalties for missing Service Standards

TBD **TBD**
TBD

Non Core Services

1. Service

TBD

Payment Terms

TBD

APPENDIX 2

PARTIES TO THE AGREEMENT

PARTIES:

1. **TBD**

2. **TBD**

<table>
<tr><td>Signed, on behalf of Business Systems Solutions by

....

Signature..
.</td><td>Signed, on behalf of TBD by

....

Signed:..
.....</td></tr>
<tr><td>CONTACTS: TBD</td><td>CONTACTS: TBD</td></tr>
<tr><td colspan="2">ARBITRATOR:</td></tr>
</table>

CUSTOMER LOCATIONS & UNITS COVERED BY THIS AGREEMENT:

APPENDIX 3

SERVICE STANDARDS REPORT

CUSTOMER: **TBD**

PERIOD OF REPORT: TBD

Service Standard	Target	Actual	Variance
TBD	TBD	TBD	TBD

PENALTIES APPLIED:

Signed..

Date..

DOCUMENT STATUS PAGE

Title	:	Service Standards between Business Systems Solutions (BSS) and **TBD**
Ref	:	**TBD**
Issue	:	1.0
Date	:	22 May xxxx
Author	:	Andrew Hiles

Authorisation

Name	Date	Signature

Change History

Issue Description	Date	Author
1.0 Initial Draft	22 May xxxx	A Hiles

Notes:

1. This document places stringent demands upon the developer. It may not be practicable to introduce all of the measurements and metrics initially: these may be phased in over time as capability and tools permit. It is thus intended to represent the destination rather than the first stop. It is intended as a discussion document and is expected to be subject to radical change following review.

2. Much of the detail contained in Model 1 is omitted: this detail may be contained in separate BSS Project Management Procedures which should include Corrective and Remedial Action and Post Implementation Review procedure and which should be referenced in Model 2 and thereby form part of Model 2, accepted by BSS and the Customer. This Manual could also contain a brief explanation of FPA or PI.

3. For the sake of brevity, some general paragraphs at Model 1 have been omitted (notably Ownership; Inspection; Fault Reporting).

4. The search string **TBD** (To Be Defined) is used to identify points at which to insert values or to amend or insert text (which may be subject to detailed consideration, discussion, negotiation and decision).

APPENDIX E:

Checklist for Outsourcing & Facilities Management

OUTSOURCING & FACILITIES MANAGEMENT: SERVICE LEVEL CRITERIA

1. The service aspects below identify some of the issues which need to be defined within a Service Level Agreement for outsourcing.

1.1 minimum operational service commitment (i.e. Terms of Service Level Agreement for operational aspects)

- hours of cover/out of hours arrangements
- reliability (maximum number of service peaks permitted)
- availability (minimum % uptime required and when)
- response

1.2 minimum support service commitment (i.e. Terms of Service Level Agreement for support aspects)

- problem management
    - priorities
    - response time by priority
    - problem reporting procedure/audience
    - escalation procedure
    - hours of cover/out of hours arrangements
    - vendor review meetings (frequency, attendance)

1.3 minimum acceptable range of equipment/applications covered

1.4 minimum acceptable vendor contingency arrangements

1.5 minimum acceptable back-up arrangements

1.6 minimum acceptable security arrangements

1.7 acceptable time slots for hardware/software routine maintenance

1.8 minimum network access/resilience requirements

1.9 acceptable policy on sub-contracting

1.10 acceptable policy on maintenance vendor (hardware/software vendor or independent maintenance vendor?)

1.11 acceptable policy on hardware replacement v maintenance

1.12 minimum notice of change impacting client interface

1.13 acceptable service quality measurement metrics and tools

1.14 provision of acceptable reporting vehicle(s) and tools to agreed audience and frequency

1.15 acceptance of financial penalty in event of service levels falling below those agreed

1.16 agreement to other key contract terms and conditions (eg arbitration)

1.17 escrow arrangement (if applicable)

1.18 confidentiality agreement

1.19 European Monetary Union compliance

2. The list above is representative rather than exhaustive and needs to be quantified and tailored to the specific needs of the outsourcing client. Some of the above items may be covered by contract: input from the Service Level Agreement Project may be required into the contract procedure.

APPENDIX F:

EXAMPLE DESKTOP SUPPORT SLA

ABC LIMITED

SERVICE LEVEL AGREEMENT

agreed between

THE ABC SYSTEMS GROUP

and

THE SYSTEMS USER GROUP

FOR DESKTOP SERVICES

For the period _______________ to ________________ inclusive

Draft 1.9

1 9 September 1999

Andrew Hiles

CONTENTS

DOCUMENT CONTROL

Document Prepared by:	A N Hiles			
Date:	19 September XXXX			
Version No:	Draft 1.9	Copy No.	of	Copies
Distribution:				

Notes:

1. XXXXX and TBD indicate specific details require consideration or completion.

SECTION 1: BASIS OF AGREEMENT

1.1 Introduction

This document expresses the service relationship between the Systems Group and the Systems User Group (SUG) on behalf of the user community in respect of Personal Computer services.

1.2 Purpose

The purpose of this document is to define the services and associated service levels to be provided by the Systems Group, together with the obligations of the user.

The methodology is as follows:

- establish business requirements and define responsibilities
- identify suitable performance measures
- develop reporting procedures which identify effectiveness by function.

This will enable the Systems Group to manage the quality of the service delivered to the end user (the Customer). at ABC House and at remote sites (including sub-contractors employed by ABC).

The services to SUG and reciprocal commitments are outlined at Section 2 and specified in detail in Section

1.3 Period of Agreement & Termination

This agreement will take effect from **XXXX** and run for one year on a rolling basis. It is renegotiable within the last three months of each 12-month period.

1.4 Variation

The agreement will require renegotiation whenever there is a significant and sustained change in the scope of the original agreement by either party. Neither party will be liable for delay or failure to perform their part of the agreement due to circumstances beyond their control.

Specifically, this SLA may need to be renegotiated in the event of significant change in service requirements or as otherwise stated below:

- should the number of connected physical devices exceed XXXX
- should the number of regular concurrent users exceed XXXX per server
- should any new networked PC application, available to more than XXXX users, be added
- should network traffic or database volumes change by more than 15%.

1.5 Responsibilities for service provision

This service is provided by the Systems Group. To deliver this service, the Systems Group depend upon Premises Management and external suppliers (e.g., maintenance vendors).

1.6 Service requirements

Service requirements are specified in the sections that follow.

1.7 Service Monitoring

The availability and reliability of the service will be monitored using statistics provided by the Systems Group Help Desk (based upon reports from users). Availability will be measured in Service Core Time (i.e. attended by support staff) and in Extra Service Time (i.e., unattended operation) over a rolling monthly period. Service hours are detailed at 2.5 below

The aim is that performance will be measured at the end-user's screen using available network monitoring tools: the appropriateness of this approach will be reviewed when experience has been gained in the use of these tools for service level monitoring.

1.8 Service Performance Review

Formal monthly Review Meetings will be arranged between the Systems Group Manager and the Systems User Group.

1.9 Key Events & Peak Periods

The Service Profile and Key events are detailed at Appendix B.

The Peak periods will be between the following times:

Daily: 0800-1000 and 1600-1700 hours
Periodic: 1 week before until 4 weeks after each Quarter Day

1.10 Definitions

1.10.1 <u>Availability and reliability</u>

PC Availability is defined as:

> "the percentage of scheduled time that each PC is functional at ABC House and at each remote site."

PC Reliability is defined as:

> "a single incident of loss of *PC availability* of service from whatever cause."

Both *PC Availability* and *PC Reliability* will be monitored and reported upon by the Systems Group to Customers via SUG over a rolling 4-week period. Availability and reliability will be measured at end users' terminals.

Service levels for *PC Availability* and *PC Reliability* are detailed at Section 3 below,

Application availability is defined as:

> "the percentage of scheduled time that a networked application is available on a server and accessible via the network to end users."

1.10.2 Response

Response is defined as the time between sending an instruction (key depression) and the time the instruction is executed and displayed on the user's screen.

Response for individual applications and PCs cannot be guaranteed since it depends on network capacity and loading and on the configuration, loading and effective disk management of the individual PC. Customers experiencing unacceptable response times should report this to the Help Desk, who will initiate an investigation and resolution of the problem.

1.11 Corporate Policy and Statutory/Regulatory Requirements

In the provision of the services contained in this agreement, both parties will ensure that Corporate policy is implemented and that the necessary checks will be put in place to ensure compliance, including statutory/regulatory compliance.

1.12 Health and Safety

It is not the Systems Group's responsibility to ensure that new equipment meets Health and Safety requirements nor to advise customers should existing equipment require changing or upgrading to meet Health and Safety requirements.

In the interests of safety, Customers will operate and maintain all computer equipment for which they are responsible properly and in accordance with supplier instructions.

Equipment installed by the Systems Group will only be installed into a suitable and safe environment but it is the responsibility of the user to ensure this environment remains in satisfactory condition. In particular, Customers

- will not block air-vents or stack items on top of computer equipment.
- will not place drinks and foreign bodies: paper clips, etc. near keyboards, monitors or other PC or network equipment

- will power off equipment if not in use and left unattended for prolonged periods.

1.13 Eye Tests

Anyone who uses a terminal or PC for more than four hours per week is entitled to an eye test with Occupational Health. Please refer to Notice Boards.

1.14 Equipment Operation: Power On/off

All electrical equipment including computer terminals, PCs, printers and other peripherals should be switched off at the device at the end of each working day unless it is needed for operational reasons to be left powered on.

The energy management system for the building shuts down the power to some floors at 2000 hours each evening and restores power at 0600 hours each morning. This action will power off any equipment not powered off by its 'owner' but is not the desired power off/on method and can cause equipment malfunction.

1.15 Equipment Cleaning

Computer equipment is cleaned regularly out of normal office hours by third party cleaners arranged by Premises. Equipment will only be cleaned if it is switched off.

Equipment will, however, be maintained in a reasonable condition by its users. Cleaning materials are also available on request from the Help Desk.

If equipment is not adequately cleaned Customers will report this to Premises.

1.16 Consumables Replenishment

Supplies of consumables are available on request from the Help Desk.

1.17 Security

1.17.1 Logical Access Protection: Password Facilities

Systems residing on PCs should be password protected with the password left personal to the User. ITP and Professional Services Paradox System are PC systems and are password protected. Simple spreadsheets and word processing documents should not be thought of as systems but may need password protection for reasons of confidentiality. However, more complicated spreadsheets and simple databases may require password protection. If in doubt, customers should seek advice from EUC.

Passwords should be at least five characters in length.

1.17.2 Protection from use of unauthorised software: Licensing of Software

All proprietary applications packages are copyright. It is illegal to have the same licensed copy on more than one PC at the same time. An inventory of all installed software is held by the Systems Group, who will conduct a periodic audit to check for unauthorised software and to check the inventory is up to date. Unauthorised software will be deleted without warning.

1.18 Backup Arrangements

To ensure that full recovery can be made following hardware problems or user error, backup of computer held data is vital: storing data on the network ensures it will be backed up by the Systems Group.

1.18.1 Backup by the Systems Group: Networked PCs

The Systems Group will back up data on servers each working evening. File space on servers for the creation of personal directories on the PC network can be made available to any networked user. This space may be used by the user to store any file that the user wants to have backed up and stored off-site. Requests for allocation of network file space must be made in writing (or by electronic mail) to the Help Desk. Users of these systems need take no further action regarding protecting this data.

1.18.2 Backup by users: Routine PC Backups

For computer systems on PCs it is the responsibility of its user to ensure that data is regularly backed up.

It is recommended that users backup all their important data onto diskettes on a regular basis. These routine backups are necessary to enable users to recover data files after mistaken deletion or after a hardware failure that results in the hard disk needing replacement.

Frequency of backups is dependent on the nature and stability of the data concerned but backups should be taken at least monthly.

Such backups are normally kept in the user's office but may be stored in a fire-safe or taken home for added protection. EUC will arrange for off-site storage of users' backups if required under individual arrangements to be agreed.

1.19 Computer Viruses

1.19.1 Virus Prevention

To prevent the introduction of computer viruses no computer software other than that supplied or authorised by the Systems Group may be installed. Standards are defined in Divisional Software Policy.

1.19.2 Virus Elimination

The Systems Group subscribes to various anti-virus software packages which will be updated regularly to the latest current stable version. This software gives instructions on how to remove or eliminate viruses. Best endeavours will be made to save data.

1.20 Archiving (Data Retention)

Both for legal and corporate reasons, data may be required to be retained for specified periods of time. It is the responsibility of the user to specify to the Systems Group any requirement for data retention beyond that required for routine backup described at 1.18 above.

SECTION 2: SERVICE DESCRIPTION

2.1 Aim of Service

The aim of the service is to provide robust Personal Computing services for ABC. Services are described in more detail at Sections 3 below.

The objective of the Systems Group is to maintain adequate capacity and resource to deliver the agreed performance targets for required volumes at agreed times. To achieve this aim, the Customer must accurately cover any factors they are responsible for which can impact the service.

Services may be made available at unscheduled times with the agreement of the Associate Director, Systems. Extensions to a service cannot be assumed as they may impact other Customers or services.

This agreement describes in Section 3 the individual services making up the total package of services provided by the Systems Group and the reciprocal commitments made by the customer.

2.2 Customers

All registered users of PCs.

2.3 Service Profile

The PC service includes hardware, software, telecommunications (including X25 across remote sites), customer support for PCs, a few residual terminals at ABC House and PCs and printers at remote sites. Services are described in more detail in Section 3.

ABC Limited uses some XXXX PCs and ~XXXX printers (standalone, switched and networked) at ABC House to deliver office support services. In normal circumstances 2 working days would be the maximum period which ABC be able to function effectively in the event of loss of PC services. Even that may be too long in relation to XXXX or payments to suppliers or at critical times of the year. (See Appendix B).

2.4 Services Included

Services covered under this SLA include:

- equipment procurement
- telecom services
- software procurement
- application development
- software version control
- user registration
- user training and advice
- access to applications
- application services

- *applications on server*

    XXXX

    - Abc1 Data

        XXXX

      E-Mail

      Project Management (PMW)

    - *User Data kept on Server*
    - *Applications on PCs*

      Microsoft Office™ XXX

      Symphonia

        (including Abc1)

      XXXX

    - printing
    - moves and installations
    - problem management
    - priorities
    - problem solving
    - escalation

2.5 Service Hours

The standard hours of access to these services (Service Core Time) are 0600 hours to 2000 hours on working weekdays. Should any user require access at other times, they should contact the Help Desk at least 3 working days in advance of the requirement.

Any PC left logged on outside these hours will be automatically disconnected. However, disconnecting in this manner does not totally clear all applications (particularly Paradox and cc:Mail), and may cause the backup to fail.

Customers requiring their PC to be able to work in stand-alone mode (ie not connected to the network) should discuss their requirements with the Help Desk.

Customers will exit networked applications before switching the PC off at the end of each day.

2.6 Public Holidays

The above service levels exclude English public holidays: the service levels at such time will be subject to agreement as the requirement arises.

2.7. Change Management

The objective of change management is to optimise service enhancement and response to new requirements while containing the risk of adverse impact on Customer Service Levels.

Except as covered in Section 3 below, at present no formal Change Control Procedure exists for PC Production Services.

2.8 Disaster Recovery and Contingency Plans

Loss of access to ABC House is covered by Corporate Business Continuity Plans. Loss of PCs and networks outside of this scenario is not at present covered.

2.9 User Development of PC Applications

2.9.1 User Development Policy

While it is policy to facilitate end user development, it is important that applications which may become mission critical are designed and tested to appropriate standards. It is the responsibility of the originating Group Manager to maintain appropriate standards. It is equally important to avoid duplication of effort. EUC should therefore be consulted before any significant development is undertaken. EUC will maintain a register of end user developments.

2.9.2 User Development: Design

Customers will consult with the Systems Group to confirm they are not duplicating data or processes already available.

2.9.3 User Development: Documentation

PC systems which in the Customer's judgement are significant to the operation of their group or the business generally should be documented sufficiently to enable colleagues to understand the purpose of the system and how it operates. Documentation should be based on an English narrative description of the objectives of the system, the information that has to be keyed in and the screen/print-outs to be produced, the validation and processing carried out by the program(s) and the principal data elements stored on disk.

2.9.4 User Development: Testing

If the system results cannot be readily checked by visual inspection customers will set up test files and carry out dummy runs to test the processing and confirm the programs process the raw data as expected.

2.10 Procurement Policy

It is the policy that all computer and communications equipment and software is purchased, leased or rented by the Systems Group. No other departmental budget will have provision for any such acquisition, except for portable PCs, which must be purchased through the Systems Group by individual departments as items of office equipment.

2.11 Standardization Policy

2.11.1 Standard Hardware

To enable EUC to be operationally effective and cost effective it is necessary to standardise wherever practical on computer hardware used.

Following analysis by the Systems Group of user requirements, appropriate hardware will be recommended, wherever possible in accordance with the Systems Group's standards.

2.11.2 Non-standard Hardware

Should the Systems Group standard hardware be inappropriate then alternative equipment may be recommended. In these cases on-going support will be considered and will if practical be done by EUC. Where this is not practical, EUC may sub-contract support but will remain responsible for it.

2.11.3 Standard Software

To enable EUC to be operationally effective and cost effective it is necessary to standardise wherever practical on the computer software used.

Standard software is detailed at 3.8.1 below.

Following analysis by the Systems Group of user requirements appropriate software will be recommended, wherever possible in accordance with the Systems Group's standards.

2.11.4 Non-standard Software

Should the Systems Group's standard software be inappropriate then alternative software may be recommended. In these cases on-going support will be considered and will if practical be done by EUC. Where this is not practical, EUC may sub-contract support but will remain responsible for SSG.

SECTION 3: SERVICES & SERVICE LEVELS

3.1 EQUIPMENT PROCUREMENT

Service Level	Customer Commitments
1. Wherever practical, equipment compliant with the Systems Group's Standards will be recommended. Should the Systems Group 's standard hardware be inappropriate then alternative hardware may be recommended. In these cases ongoing support will be considered. 2. The Systems Group will calculate costs of authorised new equipment: costings will indicate the capital costs and, for remote systems, the impact to the requesting department's systems recharge. 3. Budget cover will be assessed by the Systems Group management and proposals may be returned to the requester for financial approval. 4. Standard hardware will be made available to users within 7 working days. Timescales for provision of non-standard hardware will reflect the lead-time required by the vendor. 5. Assuming appropriate financial approval and agreement on the proposed solution and installation timescales, the equipment will be ordered; the requester will be advised of delivery; and installation plans will be agreed with the requester. 6. Installation will take place during normal working hours if possible. Where disruption to normal business would be unacceptable, installation will be scheduled outside normal working hours. This may delay the installation. 7. On completion of installation the Systems Group will update its inventory and warranty/maintenance schedules. 8. The Systems Group will support standard hardware. which will if practical be done by EUC. Where this is not practical, EUC may sub-contract support but will remain responsible for it.	1. Users will give a minimum of 10 working days notice of requirements for new equipment (excluding major moves and installations - see 3.10 below). 2. Requests for new requirements will be provided to the Help Desk from the budget holder, in writing (or by electronic mail), describing the business need rather than a technical solution. 3. SUG will prioritise requests for inclusion in the Project Catalogue. Requests endorsed by SUG will be evaluated and a technical solution (or options) will be proposed and costed. 4. Customers will give access to their work areas to permit equipment to be installed and supported.

3.2 COMMUNICATIONS SERVICES

Service Level	Customer Commitments
3.2.1 Telcom Services	3.2.1 Telcom Services
1. New Lines The Systems Group will arrange for provision of new lines and advise the cost, by telephone, to the customer.	1. Customers will give a minimum of 5 working days notice of requirements for new lines
2. Faults There will be a response to faults within 8 working hours.	2. Customers will report telephone faults to the Help Desk.
3. Configuration Changes The Systems Group will maintain configurations of telephones (e.g. Group Pickup changes)	3. Customers will give a minimum of 2 working days notice for telephone configuration changes in writing (or by electronic mail) to the Help Desk, specifying the requirement
4. Replacement Telephones The Systems Group will provide replacement handsets which will be charged at cost to the ordering customer department.	4. Customers will provide a minimum of 5 working days notice of any requirement for replacement telephones in writing (or by electronic mail) to the Help Desk, signed by the customer Group Manager.
5. Moves Moves will be undertaken in accordance with 3.10 below.	

Service Level	**Customer Commitments**
3.2.2 X.25 Services 1. New Lines The Systems Group will arrange for provision of new lines by Home Service . A minimum of 20 days notice is required for the provision of a new line. Where modems , kilostream , PADS or additional equipment are required , and items are not in stock, the lead time typically may be 4-6 weeks. In some cases Home Service may require to upgrade their capacity , which may incur additional lead time. The Systems Group will agree an implementation schedule (subject to equipment delivery) with the user. 2. Faults On working days, calls will be logged between 0800 hours and 1800 hours. There will be a response to faults within 8 working hours. 3. Configuration Changes XXXX will maintain configurations of X.25 services. 4. Moves Moves will be undertaken in accordance with 3.2.2.1 above.	3.2.2 X.25 Services 1. Customers will give the maximum practicable notice of requirements for new lines and agree an implementation schedule (subject to equipment delivery) with the Systems Group. 2. Customers will allow reasonable access by communications engineers for installation and fault fixing. 3. Customers will report faults to the Help Desk. 4. Customers will give a minimum of 20 working days notice for moves of the PC, modem or PAD, in the form of a memo to the Help Desk, specifying the requirement

3.3 SOFTWARE PROCUREMENT

Service Level	Customer Commitments
1. Requirements Analysis The Systems Group will undertake analysis of end user requirements. Wherever practicable, software compliant with the Systems Group's Standards will be recommended. Standard software is detailed at 3.8.1 below 2. Standard Software Standard software will be made available to users within 7 working days. EUC will support standard software. 3. Non-Standard Software Should the Systems Group standard software be inappropriate then alternative software may be recommended. In these cases on-going support will be considered and will if practical be done by EUC. Where this is not practical, EUC may sub-contract support but will remain responsible for it. Timescales for provision of non-standard software will reflect the lead-time required by the vendor.	1. Customers will give a minimum of 10 working days notice of requirements for new software. 2. Customers will give access to their equipment to permit software to be installed by mutual agreement. 3. Where non-standard software is acquired, SUG will prioritise requests for inclusion in the Project Catalogue. Requests endorsed by SUG will be evaluated and a technical solution (or options) will be proposed and costed.

3.4 SOFTWARE VERSION CONTROL

Service Level	Customer Commitments
1. The Systems Group will evaluate new versions of Standard Software on servers and recommend to SUG whether or not they should be implemented. 2. SUG will be consulted on proposals to upgrade non-standard software: changes of versions will not be made without SUG agreement to the change on the implementation timescale. 3. Unless exceptional circumstances prevail, no new version of software will be introduced within 6 months of its launch. 4. Lead time for changing to new versions of software will be 3 months, including evaluation. 5. The Systems Group will be responsible for ensuring all users have the same version of Standard Software. 6. The Systems Group will notify SUG of new versions of Standard Software and explain the implications of moving (or not moving) to it.	1. Requests for new versions of non-standard software will be provided to the Help Desk from the budget holder, in writing (or by electronic mail), describing the business need rather than a technical solution. 2. SUG will prioritise requests for inclusion in the Project Catalogue. Requests endorsed by SUG will be evaluated and a technical solution (or options) will be proposed and costed. 3. Customers will give access to their work areas as necessary, by mutual agreement, to permit software to be installed and supported.

3.5 APPLICATION DEVELOPMENT FOR PC SYSTEMS

Service Level	Customer Commitments
1. Requirements Analysis The Systems Group will undertake brief analysis of end user requirements, in accordance with the priority allocated by SUG, before the end user designs the specification. 2. Timescale The development timescale will be agreed with the customer and will depend on the size and complexity of the development. Typically this will be between 1 and 4 weeks. 3. Standards Applications will be developed in accordance with XXXXX 4. Support EUC will support software it develops and will keep the Specifications and code to facilitate support. Where software has been developed by the customer without EUC involvement, limited support will be available. (See also User Development at Section 2, 2.9 above).	1. Customers requiring software development will submit a request through the Help Desk. 2. SUG will prioritise requests for inclusion in the Project Catalogue. Requests endorsed by SUG will be evaluated and a technical solution (or options) will be proposed and costed. 3. After discussion with EUC, Customers will draw up a specification for the Development, which will be signed off by the customer Group Manager. 4. Customers will agree timescales with EUC. 5. Customers will give access to their equipment to permit software to be installed.

3.6 TRAINING & ADVICE

Service Level	Customer Commitments
1. <u>Training</u> Training will be provided in accordance with the schedule at 3.6.1 2. <u>Advice</u> General advice will be provided by the Help Desk. Where advice is requested in response to a user problem, advice will be provided via the Help Desk to priorities at 3. 12 below).	1. <u>Training</u> Customers will give at least 10 working days notice of training requirements. 2. <u>Advice</u> Customers requiring advice will ask via the Help Desk (see 3.12 below)

3.6.1 Training Schedule

Training Subject	Lead Time	Minimum delegates	Maximum delegates	Minimum course critique score	Location
PC initial	As agreed	1	6	8	Desk or Training Room
PC continuation	As agreed	1	6	8	Desk or Training Room
PC advanced	As agreed	1	6	8	Desk or Training Room
MS Word™	As agreed	2	6	8	Training Room
MS PowerPoint™	As agreed	2	6	8	Training Room
MS Mail™	As agreed	1	6	8	Desk or Training Room
MS Excel™	As agreed	2	6	8	Training Room
Symphonia	As agreed	2	6	8	Training Room

3.7 ACCESS TO APPLICATIONS

Service Level	Customer Commitments
1. <u>User Registration</u> 1.1 New Users The Systems Group will process valid User Registration/Change Request Forms so as to provide access to users within two working days. 1.2 Password Re-sets Passwords will be re-set, for the network (Novell) and E-mail, within 2 working hours of notification of requirement to the Help Desk. For Aztec, passwords will be re-set within one working hour of notification of requirement to the Help Desk. 2. <u>Access to Server Based Applications</u> The standard hours of access to services covered by this SLA (Service Core Time) when Applications on Server(s) are available to each user is 0600 hours to 2000 hours on working weekdays. Should any user require access at other times, they should contact the Help Desk at least 3 working days in advance of the requirement. For Help Desk and EUC Service Hours, refer to 3.12.3 below. 3. <u>Service Levels</u> The Aztec service is covered by a specific Service Level Agreement. Service Levels for applications are covered at 3.8 below.	1. <u>User Registration</u> To register as a user or to notify changes to existing users, customers will complete the User Registration/Change Request available from the Help Desk in compliance with current Registration Procedures (see Appendix D).

3.8 APPLICATION SERVICES

Service Level	Customer Commitments
1. <u>Applications on Servers</u> The applications/services identified at 3.8.1 will be provided on the server(s). Definitions of availability, reliability are at Section 1, 1.10. 1.1 Availability Standard Software, together with their associated Service Levels for Availability, are at 3.8.1 below. 1.2 Reliability There should not be more than one incident of downtime to any Group or Team within the reporting period (see 1.7 above). 1.3 Response Response for individual applications and PCs cannot be guaranteed since it depends on network capacity and loading and on the configuration (including RAM; hard disk size and performance; and clock speed), loading and effective disk management of the individual PC. Response targets therefore can only be provided across the network. Response targets are at 3.8.2 below. 2. <u>Applications on PCs</u> The application identified at 3.8.1 will be provided on PCs as required. 3. <u>Backup by the Systems Group: Networked PCs</u> The Systems Group will back up data on servers each working evening. File space on servers for the creation of personal directories on the PC network will be made available on request to any networked user.	1 Customers will promptly report any loss of service to the Help Desk. 2. Customers will follow the advice provided by the Help Desk. 3. Customers experiencing unacceptable response times should report this to the Help Desk, who will initiate an investigation and resolution of the problem. 4. Customers requiring software to be installed on their PCs will request software through the Help Desk (see also 3.3 above). Customers may also request space on a server on which to store their data. This space may be used by the user to store any file that the user wants to have backed up and stored off-site. Requests for allocation of network file space must be made in writing (or by electronic mail) to the Help Desk. Users of these systems need take no further action regarding protecting this data. (See also 1.18 above). 5. For computer systems on PCs it is the responsibility of its user to ensure that data is regularly backed up.

3.8.1 List of Standard Software and associated Service Levels

Service	Availability Target
APPLICATIONS/SERVICES ON SERVER(S)	
Abc2	99.5%
Valuations	(pending)
Abc1 Data (From Symphonia)	99.5%
Symphonia Network Version	99.5%
E- Mail Microsoft™ Mail	99.5%
APPLICATIONS ON PCs: STANDARD SOFTWARE	
MS Office™ Applications MS Word™ MS Excel™ MS PowerPoint™ MS Access™	99.5%

3.8.2 Network Response Targets

90% TARGETS

Transaction Type	Average Hour Target Response (secs)	Average Hour Actual Response (secs)	Peak Hour Target Response (secs)	Peak Hour Actual Response (secs)
Accessing application	<5		<7	
Start to Retrieve file	<3		<5	
Start to Save file	<3		<5	
Send mail to Receive Mail	<300		<420	
Open mail	<3		<5	

99% TARGETS

Transaction Type	Average Hour Target Response (secs)	Average Hour Actual Response (secs)	Peak Hour Target Response (secs)	Peak Hour Actual Response (secs)
Accessing application	<8		<12	
Start to retrieve file	<5		<7	
Start to save file	<5		<7	
Send mail to receive mail	<420		<480	
Open mail	<5		<7	

NOTE: 100% TARGETS ARE 99% TARGETS X 2

3.9 MOVES AND INSTALLATIONS

Service Level	**Customer Commitments**
It is the policy that the installation or relocation of all computer and communications hardware and software is co-ordinated by the Systems Group. 1. Installation schedule will be agreed between EUC and the requester. The Systems Group will install working equipment and applications within the agreed timescales. 2. Where other parties are involved (eg Premises Management or external suppliers/services) work schedules may be affected by these outside factors. 3. Installation will take place during normal working hours if possible. Where disruption to normal business would be unacceptable, installation will be scheduled outside normal working hours. This may delay the installation. 4. On completion of installation the Systems Group will update its inventory and warranty/maintenance schedules.	1. <u>Minor Moves</u> Requests to move, modify or reconfigure hardware must be authorised by the appropriate Group Manager and submitted in writing (or by electronic mail) to the Help Desk, with details of requirement and justification. The requester will give at least 2 week's notice for routine changes. Departmental managers are responsible for compliance with Health and Safety requirements through Premises Department. On receipt of a request it will be recorded, a request number assigned for future reference and then passed to the Help Desk for technical evaluation and/or acceptance, costing and authorisation The request will then be returned to the Help Desk who will contact the requester for final authorisation before the work is scheduled. 2. <u>Office Moves</u> For large (i.e., more than 5 devices) or wholesale departmental moves the Systems Group requires at least 4 weeks notice. It is important that the Systems Group is involved in the planning of any reorganisation and that approval by the Systems Group is given for any plans to the commitment to those plans. Departmental managers are responsible for compliance with Health and Safety requirements through Premises Department. In this way the Division can ensure that plans are technically feasible, practical, ergonomically viable, comply with Health and Safety regulations and are co-ordinated with Premises Department.

3.10 REMOTE USERS

Service Level	Customer Commitments
1. Installations at New Sites The Systems Group will acquire and install PCs and printers at remote sites.	1. Users will give a minimum of 4 weeks notice of new requirements to the Systems Group.
2. Communications Provision of communications services will be in accordance with 3.2 above.	2. Users are responsible for the provision of PC and Printer desks, stands etc and for provision of necessary power supplies to the point of installation.
3. Service Levels Except as indicated in 3.11, Service Levels outlined elsewhere in Section 3 will apply.	3. Users are responsible for Health and Safety aspects of installations and will ensure necessary checks are undertaken before installation commences. 4. Users will give reasonable access to the Systems Group and communications engineers for installation and fault fixing.

3.11 PROBLEM MANAGEMENT

Service Level	**Customer Commitments**
1. Priorities Problem priorities and escalation procedures are defined overleaf. Any problem which causes a major service or facility to be unavailable for over four working hours will result in a written report to SUG summarising the cause, resolution and actions/plans to prevent recurrence. 2. Problem Solving On receipt of a problem, the Help Desk will log it and allocate a reference number. 10% of calls will be resolved on the telephone at the time of call. 70% of problems will be resolved within target. 100% of problems will be resolved within 150% of targets defined at 3.9.1 below. The Help Desk will respond to messages left on the answer-phone by 1000 hours the following working day. 3. Escalation Escalation will be in accordance with 3.12.2 below.	1. The Help Desk should be contacted by telephoning: **000 0000**. Calls should not be made to individual members of the Systems Group EUC Team. On logging a call the caller will receive a Help Desk reference number: if the caller has not been given a number, the caller should assume the call has not been logged. Customers requiring further information about outstanding problems should contact the Help Desk and quote the Help Desk reference number.

3.12.1 PROBLEM MANAGEMENT: PRIORITIES, FEEDBACK AND RESOLUTION TARGETS

Problem Priority	Description	First Feedback to Customer	Feedback Frequency to user	Resolution Time
1	Emergency: general loss of mission-critical services	Immediate	Every working 30 minutes	Utmost speed
2	Critical: loss of mission-critical service to >10 users	30 minutes	Every working hour	30 minutes
3	Serious: loss of mission-critical service to 1-10 users	1 working hour	Every 2 working hours	1 working hour
4	Significant: loss of non-mission critical service to any user	2 working hours	Every 3 working hours	2 working hours
5	Minor: inconvenience caused; but work-around possible	1 working day	Every working day	1 working day
6	Other: cosmetic or trivial	1 working week or as agreed with user	Every working week or as agreed with user	One working week or as agreed with user

3.12.2 PROBLEM ESCALATION

Problem/Priority	Operations and Technical Support Manager	Systems Manager
Priority 1	Immediately	Immediate
Priority 2	Immediately	within 30 minutes
Priority 3	30 minutes	1 hour
Priority 4	1 hour	2 hours
Priority 5	1 day	2 days if business impact
Priority 6	1 week	1 week if business impact

3.12.3 HELP DESK AND EUC SERVICE HOURS

Service	Cover Type	Service Hours
Help Desk	attended service	Mon-Fri 0800-1800
	answer-phone service	Mon-Fri 1800-0800 Weekend 1800 Fri -0800 Mon
EUC Service	attended service	Mon-Fri 0800-1800
Production Control	attended service (core time)	Mon-Fri 0800-1800
	unattended service (extra service time)	Mon-Fri 1800-0800
On-Line systems	available with support (core time)	Mon-Fri 0730-2000
	available but no on-site attendance (extra service time)	Mon-Fri 0600-0730 Sat 0600-2000 Sun 0600-2000

Notes:

All other times, including English Public Holidays - reasonable endeavours, support by arrangement with the Help Desk.

For Aztec Service, see Aztec SLA

APPENDICES

APPENDIX A: SCHEDULE OF ITEMS COVERED UNDER MAINTENANCE

Hardware	Comments
PC	All items acquired by the Systems Group
Printers	All items acquired by the Systems Group
Servers	All items acquired by the Systems Group
Hubs	All items acquired by the Systems Group
Cabling	All items acquired by the Systems Group
Communications equipment	All items acquired by the Systems Group
Software	**Comments**

APPENDIX B: SERVICE PROFILE & KEY EVENTS

ID	Description	Frequency	Volume	Comments

APPENDIX C: SERVICE LEVEL REPORTS: ITEMS TO BE REPORTED

PERIOD ________ TO ________

C.1 PC APPLICATIONS AND SERVICES, NOVELL NETWORK

C. 1.1 Availability & Reliability

PC and Network

Service	Target No. of incidents of downtime System to end user	Actual No. of incidents of downtime System to end user	Availability Target %	Availability Actual %
Service Core Time 0600-2000 Weekdays	1 20		99.5%	
Extended Service Time (scheduled on request)	1 20		99.8%	

Printers

Service	Target No. of incidents of downtime System to end user	Actual No. of incidents of downtime System to end user	Availability Target %	Availability Actual %
Service Core Time 0600-2000 Weekdays	1 20		95%	
Extended Service Time (scheduled on request)	1 20		95%	

C.2 STANDARD SOFTWARE, SERVICE CORE TIME: SERVICE LEVEL REPORT

Service	Reliability (Number Incidents of downtime)		Availability	
	Target	Actual	Target	Actual
APPLICATIONS/SERVICES ON SERVER(S)				
Abc2	1		99.5%	
Abc3	1		(pending)	
Abc1 Data (From Symphonia)	1		99.5%	
Xyz Network Version	 11		99.5% 99.5% 99.0%	
E-Mail Microsoft™ Mail	1		99.5%	
APPLICATIONS ON PCs: STANDARD SOFTWARE				
MS Office™ Applications MS Word™ MS Excel™ MS PowerPoint™ MS Access™			99.5%	

C.3 NETWORK RESPONSE TARGETS

90% TARGETS

Transaction Type	Average Hour Target Response (secs)	Average Hour Actual Response (secs)	Peak Hour Target Response (secs)	Peak Hour Actual Response (secs)
Accessing application	<5		<7	
Start to Retrieve file	<3		<5	
Start to Save file	<3		<5	
Send mail to Receive Mail	<300		<420	
Open mail	<3		<5	

99% TARGETS

Transaction Type	Average Hour Target Response (secs)	Average Hour Actual Response (secs)	Peak Hour Target Response (secs)	Peak Hour Actual Response (secs)
Accessing application	<8		<12	
Start to retrieve file	<5		<7	
Start to save file	<5		<7	
Send mail to receive mail	<420		<480	
Open mail	<5		<7	

NOTE: 100% TARGETS ARE 99% TARGETS X 2

C.4 EUC & HELP DESK

Problems resolved on phone at time of call	Target	Actual
	10%	

Problem Priority	Target Resolution Time	95% Resolution Target Achieved	99% Resolution Target Achieved
1	30 minutes		
2	30 minutes		
3	1 working hour		
4	2 working hours		
5	1 working day		
6	1 working week or as agreed with customers		

C.5 OTHER SERVICES: EXCEPTIONS

Service	Target	Actual	Variance
Equipment Procurement			**TBD**
Software procurement			
Software Version Control			
Training & Advice			
Moves & Installations			

C.6 SERVICE OVERVIEW

C.6.1 Members of the System User Forum are requested to complete the following summary each month in time for Service Review Meetings. The form should also be used as a means to review the service at Group and Team meetings. The rating column enables easy usage and common understanding of the "softer" aspects of the service.

<u>Service Rating Explanation</u>

10. Excellent
9. Good
8. Satisfactory
6. Minor Production impact causing inconvenience
4. Major Production impact which is unacceptable
2. Very bad service
0. No service at all

Service	Customer Rating	Target	Weight Factor	Comments

C.6.2 Service Rating

The Service Rating can be calculated by taking into account the weighting factors when considered against targets.

C.6.3 Service Management

The Systems Group will maintain details and statistical information on all services objectively to consider the cause of a low rating so as to improve service.

This service will be reviewed monthly following production of the Service Level Reports.

APPENDIX D: PROCEDURES FOR REGISTERING/AMENDING USER ACCESS

[Contents Omitted]

APPENDIX E: PARTIES TO THE AGREEMENT AND CONTACTS

PARTIES:

1. **The Systems Group**
2. **The Systems User Group**

Signed, on behalf of the Systems Group by	**Signed, on behalf of the SUG by**
.... Signature..	.... Signed:.. ...
CONTACTS: TBD TEL:	**CONTACTS: TBD TEL:**

BIBLIOGRAPHY

Barham, W. J. & Murrey Faithfull, "Service Level Management at Lloyds Bank", Conference Proceedings of Computer Measurement Group (1988), pp 577-582.

Clifford-Winters, Tony, "Service Level Agreements: To Negotiate or Impose", Conference Proceedings of UK Computer Measurement Group (May 1986), pp 37-59.

Dugmore, Jenny Dr., "Managing the Gap Between Service Targets and Actual", Conference Proceedings of Computer Measurement Group (May 1990), pp 544-551.

Dugmore, Jenny Dr., "Service Level Management . A Pragmatic Approach", Conference Proceedings of Computer Measurement Group (June 1989), pp 259-264 and Conference Proceedings of SHARE Europe (April 1990), pp 55 1-522.

Elliot, Tim., "Pitfalls and Joys of Service Level Agreements", Conference Proceedings of SHARE Europe (April 1990), pp 50 1-509.

Hiles, Andrew N., "Service Level Agreements: Managing Cost and Quality in Service Relationships", Chapman & Hall, 1993, ISBN 0 412 54240 4

Hiles, Andrew N. *E-Business Service Level Agreements: Strategies for Service Providers, E-Commerce and Outsourcing, 2002.* Published by Rothstein Associates Inc. , ISBN #0-9641648-9-2.

Howard, Phillip C., "Standard Costing in Data Processing", EDP Performance Review (June 1981), pp 1-6.

Miller, George W. A. (Bill), American Airlines: "Service Level Agreements: Good Fences Makes Good Neighbours", Conference Proceedings of Computer Measurement Group (1988), pp Miller 1-2 1.

Propst, John W., "Developing Effective Service Level Agreements," Conference Proceedings of Computer Measurement Group (December 1985), pp 704-709.

Rene, Jacques, "Service Level Agreements: Practical Implementation" Conference Proceedings of Computer Measurement Group (May 1990).

Sherkow, Alan M, "Using Service Level Objectives", Conference Proceedings of Computer Measurement Group (December 1986), pp 623-628.

Wenk, Dennis, "Management Control of Management Information Systems", Conference Proceedings of Computer Measurement Group (December 1985), pp 486-489.

ABOUT THE AUTHOR

ANDREW HILES was the founding chairperson of EURIM, the working group supporting the UK all-party Parliamentary Working Party on European IT legislation. He is founder and Chairman of Survive, the international user group for business continuity planning, and was a founding Director of the Business Continuity Institute, the international body for certification of business continuity professionals. He is a founder Director of Kingswell, international consultants. Having commenced his management career with the Royal Air Force, he pioneered IT systems before leaving to take up a position within the Finance Department of London Transport. Subsequently in their Central Productivity Unit he was a Senior Projects Manager and later became responsible for the business re-engineering function, implementing new services and major technical projects. He left to take up a position with the UK Post Office as their first Business Systems Consultant responsible for major projects. Andrew then joined the UK Atomic Energy Authority at the Harwell Laboratories where he managed the supercomputing, mainframe and other bureau and outsourcing services. Andrew is a pragmatic consultant and trainer in the areas of Business Continuity Planning and Service Management

Andrew left Harwell to set up Kingswell, an international training and consulting company specializing in service management, customer – supplier relationships and enterprise risk management.

He has helped hi-tech, financial, transport and government bodies to develop and enhance Service Management, Customer Support and Service Desk functions and has supported both customers and suppliers in Service Level Agreements, Market Testing and Outsourcing.

Andrew is an international speaker on service management and has featured on conference programs in the USA, Southern Africa, Europe, the Middle East and the Pacific Rim. He has presented workshops and seminars on these topics for Frost & Sullivan (Europe), IIR/ IFF (Europe and Middle East), AIC (South Africa), CEL (Hong Kong), UPOM (Middle East) and other companies having also lectured at Ashridge, Cranfield, GEC Dunchurch and Henley Management Colleges in the UK. He has broadcast on TV, radio and Internet webinars.

He has over 300 published articles on service management and is author of two other books on Service Level Agreements.

Andrew is a Fellow of the Business Continuity Institute, a Member of the British Computer Society and a Freeman of the City of London.

OTHER SLA TOOLS AND RESOURCES BY ANDREW HILES

SLA FRAMEWORK™: SERVICE LEVEL AGREEMENT FRAMEWORK ON CD-ROM

SLA FRAMEWORK™ brings together the critical elements needed to build a Service Level Agreement. All you do is choose the plan elements you require, load them into a standard word processor, edit them to your specifications, and you're done! It's that simple. No programming experience required. SLA FRAMEWORK™ is easy to operate. It will save you days, weeks, possibly even months of valuable time. Now every IT Service professional can have effective SLAs!

2001, CD-ROM. ISBN #0-9641648-5-X

SERVICE LEVEL AGREEMENTS: Winning A Competitive Edge for Support & Supply Services

This book holds the key to creating enduring, satisfying and profitable relationships between customer and supplier. It shows how both internal and external services and supply can be aligned to meet business vision, mission, goals, critical success factors and key performance indicators. The techniques described will help you balance service cost against quality, leading to competitive advantage and business success. They can be applied to any industry, to any supply or support service. They have been used by leading companies internationally - and they work!

2000, 296 pages (approx.). ISBN #0-9641648-4-1

OTHER SLA TOOLS AND RESOURCES BY ANDREW HILES (Continued)

E-Business Service Level Agreements: Strategies for Service Providers, E-Commerce and Outsourcing

This book by Andrew Hiles, an internationally respected consultant, author and educator, provides important guidelines and insights about SLAs and how they factor into e-business. The book is supported by anecdotal information, industry statistics, and product descriptions. One can actually get a solid picture of how an SLA ought to fit into the grand scheme of things.

2002, 180 pages (approx.), ISBN 0-9641648-9-2

Mr. Hiles is also the author of **Business Continuity: Best Practices** (2001, ISBN #0-9641648-3-3), and co-author with Dr. Y. Gunn, **Creating a Customer-Focused Help Desk: How to Win and Keep Your Customers** (2000, ISBN 0-9641648-6-8). Published by Rothstein Associates Inc.

These and hundreds of other books, software tools, videos and research reports are available from Rothstein Associates Inc.

The Rothstein Catalog on Service Level Books
www.ServiceLevelBooks.com

The Rothstein Catalog On Disaster Recovery
www.DisasterRecoveryBooks.com

Rothstein Associates Inc.
4 Arapaho Road, Brookfield Connecticut 0608-3104 USA
203.740.7444 or 1-888-ROTHSTEin; fax 203.740.7401
info@rothstein.com

THE COMPLETE GUIDE TO I.T. SERVICE LEVEL AGREEMENTS:

ALIGNING IT SERVICES TO BUSINESS NEEDS (3RD ED.)

by Andrew Hiles, ISBN #1-931332-13-4

PRODUCT REGISTRATION AND REQUEST FOR FREE CD-ROM!

Purchasers of **The Complete Guide to IT Service Level Agreements (3rd Ed.)** are entitled to a ***FREE*** **CD** containing **THE ROTHSTEIN CATALOG ON DISASTER RECOVERY!**

If you purchased this book other than directly from Rothstein Associates Inc., email or fax the information below for your **FREE** CD. *(If you purchased this book directly from Rothstein Associates Inc.– The Rothstein Catalog On Service Level Books or The Rothstein Catalog On Disaster Recovery– you are automatically registered, and the CD-ROM is packaged with your book).*

*To register your purchase and to receive your **FREE** CD-ROM, fax the following information to 203.740.7401, or email to info@rothstein.com:*

Product: **The Complete Guide to IT Service Level Agreements (3rd Ed.)**

First Name ______________ Last Name ______________

Company/Organization ______________

Dept./Mail Station ______________ Title ______________

Street Address ______________

City ______________ State/Province ______________

Zip/Postal Code ______________ Country ______________

Email address ______________ Phone ______________

Where Purchased ______________ Purchase Date ______________

Check here if you would like to receive a complimentary CD-ROM containing **THE ROTHSTEIN CATALOG ON DISASTER RECOVERY**, the industry's principal resource for hundreds of books, software tools, videos and research reports since 1989. ☐

Enter your email address here if you would you like to receive a complimentary subscription to our email newsletter, **BUSINESS SURVIVAL**™: Business Continuity and Service Level Agreements for Key Decision-Makers

____

THE ROTHSTEIN CATALOG ON SERVICE LEVEL BOOKS

ROTHSTEIN ASSOCIATES INC.

4 Arapaho Rd., Brookfield, Connecticut 06804-3104 USA

203.740.7444 Fax 203.740.7401

info@rothstein.com

www.rothstein.com or **www.ServiceLevelBooks.com**

www.ingramcontent.com/pod-product-compliance
Lightning Source LLC
LaVergne TN
LVHW080033210926
848481LV00011B/3178